W0192933

Topics in Current Physics

25

# Topics in Current Physics Founded by Helmut K. V. Lotsch

1 **Beam-Foil Spectroscopy**
Editor: S. Bashkin

2 **Modern Three-Hadron Physics**
Editor: A. W. Thomas

3 **Dynamics of Solids and Liquids by Neutron Scattering**
Editors: S. W. Lovesey and T. Springer

4 **Electron Spectroscopy for Surface Analysis**
Editor: H. Ibach

5 **Structure and Collisions of Ions and Atoms**
Editor: I. A. Sellin

6 **Neutron Diffraction**
Editor: H. Dachs

7 **Monte Carlo Methods** in Statistical Physics
Editor: K. Binder

8 **Ocean Acoustics**
Editor: J. A. DeSanto

9 **Inverse Source Problems** in Optics
Editor: H. P. Baltes

10 **Synchrotron Radiation**
Techniques and Applications
Editor: C. Kunz

11 **Raman Spectroscopy** of Gases and Liquids
Editor: A. Weber

12 **Positrons in Solids**
Editor: P. Hautojärvi

13 **Computer Processing of Electron Microscope Images**
Editor: P. W. Hawkes

14 **Excitons** Editor: K. Cho

15 Physics of **Superionic Conductors**
Editor: M. B. Salamon

16 **Aerosol Microphysics I**
Particle Interactions
Editor: W. H. Marlow

17 **Solitons**
Editors: R. Bullough, P. Caudrey

18 **Magnetic Electron Lens Properties**
Editor: P. W. Hawkes

19 Theory of **Chemisorption**
Editor: J. R. Smith

20 **Inverse Scattering Problems** in Optics
Editor: H. P. Baltes

21 **Coherent Nonlinear Optics**
Recent Advances
Editors: M. S. Feld and V. S. Letokhov

22 **Electromagnetic Theory of Gratings**
Editor: R. Petit

23 **Structural Phase Transitions I**
Editors: K. A. Müller and H. Thomas

24 **Amorphous Solids**
Low-Temperature Properties
Editor: W. A. Phillips

25 **Mössbauer Spectroscopy II**
The Exotic Side of the Method
Editor: U. Gonser

26 **Crystal Cohesion and Conformational Energies**
Editor: R. M. Metzger

27 **Superfluorescence, Resonance Fluorescence and Optical Bistability**
Editor: R. Bonifacio

# Mössbauer Spectroscopy II

## The Exotic Side of the Method

Edited by U. Gonser

With Contributions by
R. L. Cohen   H. Fischer   V. I. Goldanskii
U. Gonser   S. S. Hanna   W. Hoppe
R. N. Kuzmin   R. L. Mössbauer   V. A. Namiot
F. Parak   R. V. Pound   R. S. Preston
B. D. Sawicka   J. A. Sawicki

With 67 Figures

Springer-Verlag Berlin Heidelberg GmbH 1981

Professor Dr. Ulrich Gonser
Universität des Saarlandes, Fachbereich Angewandte Physik,
D-6600 Saarbrücken, Fed. Rep. of Germany

ISBN 978-3-662-08869-2 ISBN 978-3-662-08867-8 (eBook)
DOI 10.1007/978-3-662-08867-8

Library of Congress Cataloging in Publication Data. Main entry under title: Mössbauer spectroscopy II. (Topics in current physics ; 25). Bibliography: p. Includes index. 1. Mössbauer spectroscopy. I. Gonser, U. II. Cohen, Richard Lewis, 1936– . III. Series. QC491.M62 537.5'352 81-4830 AACR2


Originally published by Springer-Verlag Berlin Heidelberg New York in 1981.

Softcover reprint of the hardcover 1st edition 1981



2153/3130-543210

# Foreword

Some newly discovered effects lose their glamor after a short period of euphoria. Others, however, retain their fascination for a long time and, even as they mature, display unexpected features. The Mössbauer effect belongs to the second category. Rudolf Mössbauer's discovery of recoilless gamma-ray emission in 1957 immediately caused a flurry of attention, and confirming work appeared almost at once. Since then the flow of publications has steadily increased. Most studies follow predictable paths; the essential aspects of these "conventional" experiments have been described in the first volume of the present work (*Mössbauer Spectroscopy*, Topics in Applied Physics, Vol. 5). These straightforward investigations have not, however, exhausted the field, boredom has not set in, and unexpected applications continue to appear. In the present volume, Uli Gonser has collected contributions that display the "exotic" side of the Mössbauer effect. They range from a masterly description of the red-shift experiment to a clear exposition of a powerful solution to the old and painful phase problem in crystallography. Each of the contributions exhibits a different side of recoilless gamma-ray emission. Together they show that the field is very much alive and continues to delight us with elegant solutions to old problems, unanticipated glimpses at new phenomena, clever uses of new technical possibilities, and ingenious applications to fields far away from physics. I believe that novel features of the Mössbauer effect will continue to appear and that new applications will still be found. The contributions to the present volume provide inspiration and guidance in the search for new frontiers.

Urbana, Illinois *Hans Frauenfelder*
May 1981

## Contents

## List of Contributors

Cohen, Richard L.

Bell Laboratories, Murray Hill, NJ 07974, USA

Fischer, Harald

Universität des Saarlandes, Fachbereich Angewandte Physik, D-6600 Saarbrücken, Fed. Rep. of Germany

Goldanskii, Vitalii I.

Institute of Chemical Physics, Vorobjevskoje Shosse 2-B, Moscow 117334, USSR

Gonser, Ulrich

Universität des Saarlandes, Fachbereich Angewandte Physik, D-6600 Saarbrücken, Fed. Rep. of Germany

Hanna, Stanley S.

Stanford University, Department of Physics, Stanford, CA 94305, USA

Hoppe, Walter

Max-Planck-Institut für Biochemie, Abteilung für Strukturforschung I, D-8033 Martinsried, Fed. Rep. of Germany

Kuzmin, Runar N.

Physical Faculty, Moscow State University, Leninskiye Gory, Moscow, 117234, USSR

Mössbauer, Rudolf L.

Technische Universität München, Physik Department, James Franck-Straße, D-8046 Garching, Fed. Rep. of Germany

Namiot, Vladimir A.

Nuclear Physics Research Institute, Moscow State University, Leninskiye Gory, Moscow, 117234, USSR

Parak, Fritz

Technische Universität München, Physik Department, James Franck-Straße, D-8046 Garching und Max-Planck-Institut für Biochemie, D-8033 Martinsried, Fed. Rep. of Germany

Pound, Robert V.

Harvard University, Lyman Laboratory of Physics,
Cambridge, MA 02138, USA

Preston, Richard S.

Northern Illinois University, Department of Physics,
De Kalb, IL 60115, USA

Sawicka, Barbara D.

Institute of Nuclear Physics,
31-342 Cracow, Poland

Sawicki, Jerzy A.

Institute of Nuclear Physics,
31-342 Cracow, Poland

# 1. Introduction

U. Gonser

With 1 Figure

About a quarter of a century ago Rudolf Mössbauer discovered [1.1,2] that the emission and absorption of $\gamma$-rays can occur in a recoil-free fashion. This observation developed into an important new method: Mössbauer spectroscopy. This method has now reached the age of maturity as is best demonstrated by the wealth of information which has been obtained in all disciplines of the natural sciences. Also, Mössbauer resonance has been observed in more than 100 isotopes and about 120 excited states, and the number of published papers dealing with this method has reached the order of $10^4$.

The former Topics volume [1.3], published in 1975, was concerned with the "classical" approach; that is, it described the usefulness of the method when applied in a "classical fashion" to the "classical disciplines", such as chemistry, magnetism, biology, geology and mineralogy, and physical metallurgy.

The phrase "classical fashion" indicates that most of the results were obtained by the common arrangement with the four basic components: single-line source, absorber, drive system – moving the source relative to the absorber whereby a Doppler modulation of the $\gamma$-ray energy occurs – and detector. This is shown schematically in Fig.1.1. Source and absorber are represented in their excited and ground nuclear states. The nuclear resonance transition is indicated by the bold arrows connecting the two states.

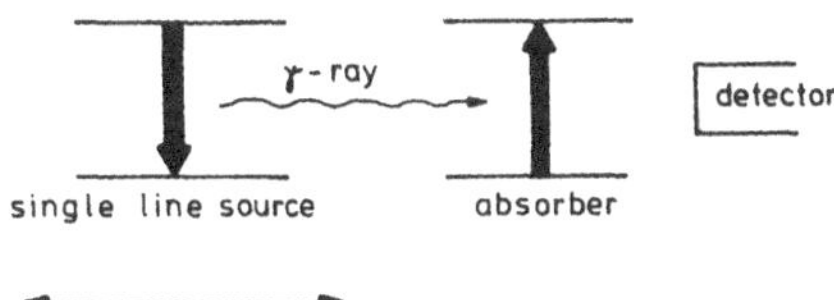


Fig. 1.1

By contrast, the present book illustrates the sophisticated side of the Mössbauer effect, Here it is our intention to demonstrate the wide variety and the great potential of $\gamma$ resonance beyond what is commonly meant by the "Mössbauer Effect". In fact, some of the ideas and applications described are so extraordinary that the word "exotic" might be especially appropriate. The experimental set-ups are different, some of which are now being tested in laboratories, while others are still purely "Gedankenexperimente" which might be carried out at some time in the future.

Table 1.1

| Chapters \ "Classical" Mössbauer components | Source | Absorber (scatterer) (transmitter) | "Drive system" (modulation) | Detector |
|---|---|---|---|---|
| 2. Phase problem | strength (≈ Ci) | γ-resonance scatterer | | large area position-sensitive proportional counter |
| 3. Gravitational red shift | large separation between source and absorber<br>precision shift measurements | | gravitational potential | |
| 4. γ-laser (gaser) | populating (pumping) nuclear excited levels. stimulated emission of Mössbauer γ-rays | resonator | positive feedback | |
| 5. synchrotron radiation | synchrotron radiation | nuclear excitation, Bragg scattering | monochromator | gated and delayed techniques |
| 6. polarimetry | polarized γ-rays (polarizer) | polarized γ-rays (analyzer) birefringence transmitter | polarization | |
| 7. implantation, conversion electron Mössbauer spectroscopy CEMS | | implantation of resonance nuclei, emitted electrons | | electron counter |

In Table 1.1 we have attempted to indicate the peculiarities in the methodology of the following six chapters. Some of the basic components of the classical set-up, as seen in Fig.1.1, are still present or are recognizable, however some of them have undergone significant modifications.

The Mössbauer effect has led to contributions in many fields. Some of the applications deserve adjectives such as funny, sophisticated, strange or surprising. Such experiments are described in Chap.8. Finally, some historical facts from the exciting early days of the discovery when people started to believe in the existence of the Mössbauer effect are related in Chap.9.

Of course, in this rather diversified collection of contributions each chapter has to stand by itself, assuming familiarity with the basic principle [1.3]. Nevertheless, some correlations can be found. For instance, if one believes the optimists in the field of synchrotron radiation, we are on the threshold of a new important tool with wide applicability. Thus, the phase problems discussed in Chap.2 as well as the gravitational red shift treated in Chap.3 might be tackled with greater accuracy using synchrotron radiation.

This Topics volume – like the previous one [1.3] – is addressed not particularly to the experts but rather to persons interested in learning what has been done, what can be done and what might be done with $\gamma$-resonance methods. Naturally questions are often formulated instead of answers. However, one can predict that in the 80's many of these problems will find their solutions. This book outlining typical examples of interdisciplinary research should transmit the spirit of frontiers in science.

These contributions can be seen as partial fulfillment of Mössbauer's dream expressed in the last sentence of his Nobel laureate address in 1961 when he said that the effect bearing his name "makes possible new advances in the exciting world of unknown phenomena and effects".

## References

1.1 R.L. Mössbauer: Z. Physik *151*, 124 (1958)
1.2 R.L. Mössbauer: Naturwissenschaften *45*, 538 (1958)
1.3 U. Gonser (ed.): *Mössbauer Spectroscopy*, Topics in Applied Physics, Vol.5 (Springer, Berlin, Heidelberg, New York 1975)

# 2. A Solution of the Phase Problem in the Structure Determination of Biological Macromolecules

R. L. Mössbauer, F. Parak, and W. Hoppe

With 12 Figures

The structure determination of biological macromolecules by single-crystal diffraction techniques requires an experimental solution of the phase problem inherent in the structure analysis. This chapter describes a method for the phase determination which uses the nuclear resonance scattering of an $^{57}$Fe nucleus as reference. The basic features of the method are discussed and different techniques are compared. The pertinent amplitude for nuclear resonance scattering is derived. First test experiments on single crystals of myoglobin are described. Technical problems arise from the necessity of using gamma-ray sources with limited intensities. In particular, the optimization of $^{57}$Co sources, the development of large-area position-sensitive proportional counters, and the necessary freezing of protein crystals are discussed. The measuring time required for a structure determination of bacterial catalase is estimated.

## 2.1 Introductory Comments

An understanding of the biological activity of proteins depends on a prior understanding of their structure. X-ray diffraction analysis in recent years has substantially increased the structural information on biological macromolecules, and the understanding of biological processes at the molecular level, by consequence, has grown vigorously. More and more studies of the active centers of biological macromolecules are also performed by means of spectroscopic techniques. A fertile application of such methods usually depends likewise on a preceding structure analysis.

Structure determinations of proteins at present are almost exclusively performed by means of X-ray diffraction measurements on protein single crystals. Such structure determinations rest on intensity measurements and experimental phase determinations for the waves scattered in the various possible Bragg directions. The supplementary phase determinations almost always employ the method of "multiple isomorphous replacement" developed by PERUTZ and KENDREW [2.1-4]. This method shows limitations with respect to the molecular weight of the systems which can be studied. Table 2.1 gives examples of proteins with differing molecular weights. Immunoglobulin-gamma

Table 2.1. Molecular weights of several proteins

| Protein | Molecular weight [g/mol] |
|---|---|
| Ferredoxin | 6 000 |
| Myoglobin | 17 000 |
| Hemoglobin | 65 000 |
| Immunoglobulin-γ | 150 000 |
| Catalase | 240 000 |
| Tropocollagen | 360 000 |
| Hemocyanin | 8 900 000 |
| Tabacco mosaic virus | 39 400 000 |

with $M = 150000$ g/mol represents the typical upper limit for structure determinations by means of multiple isomorphous replacements. Beyond that limit, new experimental techniques are required. One may, in particular, resort to resonance scattering techniques. One technique of this kind uses the recoilless resonance scattering of gamma radiation (Mössbauer effect) from nuclei of $^{57}$Fe. This method employs the interference between gamma radiation scattered by the electrons of all the atoms (Rayleigh scattering) in a crystal and by some nuclei of $^{57}$Fe implanted at specific locations in the unit cell (Mössbauer scattering).

We mention in passing another potential application of the Mössbauer effect for structure determinations of biomolecules. The X-ray diffraction from proteins necessitates the employment of single crystals, but the formation of protein crystals in many cases proves to be difficult, if not impossible. However, since heavy macromolecules are often composed of smaller subunits, a knowledge of the assemblage of the subunits ("quarternery structure") may often suffice. Such a knowledge can in principle be derived from small-angle X-ray scattering studies of heavy-atom-labeled biopolymers in a noncrystallized state [2.5]. PARAK et al. [2.6] have discussed a possible determination of quarternery structures by means of the Mössbauer effect.

The determination of the phases of diffracted waves by means of gamma resonance scattering was suggested quite early [2.7-12] and various preliminary studies were performed on crystals with rather small unit cells [2.13-17]. This paper is concerned with potential applications of the gamma resonance method to crystals with very large unit cells, in particular aiming at structure determinations of proteins with $M \geq 100000$ g/mol. Previous accounts of this method have appeared elsewhere [2.18-21].

Section 2.2 outlines the general features of the phase problem in the structure analysis of macromolecular systems. Section 2.3 gives a discussion of the experimental techniques and reviews the present state of the art. Section 2.4 describes the potential application of the effect to the solution of a protein structure with large molecular weight.

## 2.2 The Phase Problem in the Structure Analysis of Macromolecules and Its Experimental Solutions

### 2.2.1 Basic Aspects

The typical experimental arrangement of a diffraction experiment is shown in Fig.2.1. X-ray diffraction techniques have been widely used in the determination of the position coordinates of atoms, or more generally, electron-charge-density distributions within the crystallographic unit cell. One thereby measures the diffraction of X rays with wavelengths around 1 Å by a single crystal, the unit cell of which contains one or more molecules of the substance under study. Constructive interference between the waves scattered from the periodic assembly of all cells in the single crystal gives rise to intensity maxima in certain spatial directions. These special directions, which follow from the Laue equations, are characterized by their Miller indices ($hk\ell$). The intensity maxima are called the Bragg reflections, since one may consider each diffraction peak as originating from a reflection by a set of parallel lattice planes where the reflection angle $\vartheta = \vartheta(hk\ell)$ is called the Bragg angle. The spatial orientations of the various Bragg reflections relative to the direction of the incident beam (compare Fig.2.1) are exclusively determined by the geometry of the crystallographic cell, i.e., by the three basis vectors which define the unit cell of the crystal. They are independent of the kind and spatial arrangement of the atoms within the unit cell. This arrangement within the cell, however, determines the intensities of the Bragg reflections. The various Bragg reflections are thus determined with respect to their directions by the external geometric form of the unit cell, and with respect to their intensities by the internal structure of this cell.

We now introduce the electronic density $\rho(\underline{r})$ and the wave vectors $\underline{k}_1$ and $\underline{k}_2$ of the incident and scattered radiation, respectively, which are illustrated in Fig.2.2.

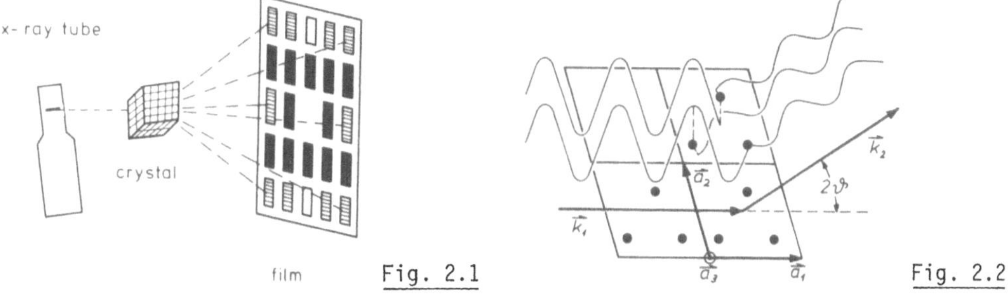


Fig. 2.1. Scheme of a diffraction experiment of X rays on a single crystal. The arrangement of the Bragg peaks on the film is determined by the space group of the crystal while their intensities reflect the structure of the molecules contained in one unit cell. The $K_\alpha$-radiation of Cu ($\lambda$=1.54 Å) or Mo ($\lambda$=0.71 Å) X-ray tubes is frequently used

Fig. 2.2. Scattering on a crystal, which is symbolized by 4 unit cells. Three equal atoms are responsible for the electronic density $\rho(\underline{r})$ of the unit cell (compare the text)

The resultant scattering amplitude of the radiation scattered from the entire unit cell, which depends on the scattering angle $2\theta$, is apparently proportional to a quantity $F(hk\ell)$, which is defined by

$$F(hk\ell) = \int \rho(\underline{r}) \exp[i(\underline{k}_2-\underline{k}_1)\underline{r}]d\underline{r} \quad , \qquad (2.1)$$

where the integration extends over the crystallographic unit cell. This equation simply represents the addition of the amplitudes of the scattered waves, with their proper phases, from each charge element. Equation (2.1) simplifies in the Bragg directions. We introduce for this purpose the basis vectors $\underline{a}_1$, $\underline{a}_2$, $\underline{a}_3$ of the unit cell of the lattice together with the basis vectors $\underline{a}_1^*$, $\underline{a}_2^*$, $\underline{a}_3^*$ of the reciprocal lattice. We also introduce the fractional atomic position coordinates within the unit cell, X, Y, Z, which are defined by the relation

$$\underline{r} = X\underline{a}_1 + Y\underline{a}_2 + Z\underline{a}_3 \quad .$$

Using the Bragg equation

$$\underline{k}_2 - \underline{k}_1 = 2\pi(h\underline{a}_1^*+k\underline{a}_2^*+\ell\underline{a}_3^*)$$

together with the usual reciprocity relations $a_i a_j^* = \delta_{ij}$ yields for the wave scattered by the unit cell in a Bragg direction according to (2.1)

$$F(hk\ell) = \int \rho(XYZ) \exp[2\pi i(hX+kY+\ell Z)]dXdYdZ \quad . \qquad (2.2)$$

The integration extends again over the unit cell. The quantity $F(hk\ell)$, which is proportional to the amplitude of the scattered wave, is conventionally called the structure factor, because it reflects the electronic density distribution and therefore the atomic structure within the unit cell. The periodic arrangement of the unit cells within the crystal allows a solution of (2.2) for the electronic charge density by means of a Fourier-sum inversion:

$$\rho(XYZ) = (V)^{-1} \sum_{hk\ell} F(hk\ell) \exp[-2\pi i(hX+kY+\ell Z)] \quad , \qquad (2.3)$$

where V is the volume of the unit cell. Actual applications use only a limited number of indices $(hk\ell)$ in (2.3) as a consequence of a limited number of measurements of reflection intensities. This limitation restricts the achievable spatial resolution, i.e., the size of the elements $\rho(XYZ)$ between which the density $\rho(XYZ)$ and $\rho(X+\Delta X\ Y+\Delta Y\ Z+\Delta Z)$ can still be distinguished.

The determination of the electronic charge density, which peaks at atomic locations, according to (2.3) requires a complete knowledge of the structure amplitude $F(hk\ell)$:

$$F(hk\ell) = |F(hk\ell)| \exp[i\phi(hk\ell)] \quad . \tag{2.4}$$

Intensity measurements in the various Bragg directions, being proportional to $|F(hk\ell)|^2$, yield only the module of the structure factor $F(hk\ell)$, while the information on the phase $\phi(hk\ell)$ according to (2.4) gets lost. A complete structure determination based on the use of (2.3) therefore requires the intensity measurements to be supplemented by independent determinations of the phases $\phi(hk\ell)$ for the various Bragg reflections. This is the well-known phase problem of structure analysis, which for sufficiently small sizes of the unit cells can readily be eluded by various methods. The large molecular systems discussed in this paper, by contrast, make experimental determinations of the phases indispensable.

Phase determinations on macromolecular systems in essence involve the introduction of a reference scatterer into the crystallographic unit cell. This technique has been used to solve the structures of all the proteins which sofar have been successfully analyzed. One thereby combines the intensities $I_P = I_P(hk\ell)$ measured for a native protein crystal in the various Bragg reflections with intensities $I_{P+R} = I_{P+R}(hk\ell)$ which are obtained if one or several reference scatterers have been implanted into specific positions inside the unit cell. Specifying the scattering factor for the scattering from the entire native cell by $F_P = F_P(hk\ell)$ and that for the scattering from the implanted reference scatterer by $F_R = F_R(hk\ell)$, we obtain for the scattered intensities from the two crystals, respectively:

$$I_P = C\Big|\,|F_P|\exp(i\phi_P)\Big|^2 = C|F_P|^2 \quad , \tag{2.5}$$

$$\begin{aligned} I_{P+R} &= C\Big|\,|F_P|\exp(i\phi_P)+|F_R|\exp(i\phi_R)\Big|^2 \quad , \\ &= C\Big[\,|F_P|^2+|F_R|^2+2|F_P|\,|F_R|\cos(\phi_P-\phi_R)\Big] \quad . \end{aligned} \tag{2.6}$$

The proportionality factor C can be deduced from the experimental data by a calibration procedure like the Wilson plot. The quantity $|F_P(hk\ell)|$ for the unit cell of the native protein results from a measurement according to (2.5).

The reference scatterer has to occupy the same position in each elementary cell. It often proves difficult to introduce a single reference scatterer into the unit cell. In the general case, $F_R$ represents the vector sum of the scattering amplitudes of all the reference scatterers in the unit cell. A determination of the positions of the different reference scatterers in the unit cell, performed by means of a Patterson analysis, must be carried out. It is then straightforward to evaluate for the assembly of reference scatterers in the unit cell $|F_R(hk\ell)|$ and the phase $\phi_R(hk\ell)$ which enter in (2.6). A measurement of the intensity $I_{P+R}$ then yields two different values for the unknown phase $\phi_P$, because this phase appears in (2.6) as the argument of a trigonometric function. This two-fold ambiguity can be removed by repeating the measurement with still another reference scatterer implanted into the native unit cell.

The standard method of phase determination employed in protein crystallography is the multiple isomorphous replacement based on the use of heavy-atom reference scatterers. This method, which has been developed by PERUTZ and KENDREW [2.1-4], involves the introduction of heavy atoms, frequently mercury or uranium, into the elementary cell of crystalline protein. The intensities of a large number of Bragg reflections are measured for the native crystal and for the heavy-atom derivatives. Though this method has proven most successful, there still exists much interest in supplementary methods. This in part is due to the difficulties which arise out of the necessity to implant the heavy atom without disturbing otherwise the unit cell. Another difficulty appears for very heavy unit cells. At the average of a large number of Bragg reflections the relative size of the scattering amplitude $|F_R|$ of a heavy atom as compared to the total amplitude $|F_P|$ of all other atoms in the native cell of the crystalline protein decreases with increasing size of the unit cell. The last term in (2.6) then decreases relative to the first term, which grows proportional to the molecular weight of the protein. By consequence, the method of the isomorphous replacement of heavy atoms decreases in phase sensitivity as one advances towards very large molecules. The intensity contrast in such situations might be improved by employing clusters of heavy atoms, such as for instance several mercury atoms. Such a procedure would immediately improve the intensity contrast produced by the heavy atoms, but it appears doubtful whether isomorphous replacements can be achieved. An alternative would be the chemical decomposition of biopolymers into smaller subunits or monomers. Structure determinations of such subunits are more easily performed. Under correct physical-chemical conditions these subunits come together to form quite automatically the higher structure of the polymer, and this structure can now be determined without having recourse to experimental phase determinations. This procedure necessitates an extensive preparatory chemistry. Systematic studies of this method are not yet available.

The method of multiple isomorphous replacement employs the electronic Rayleigh scattering from heavy atoms. One may in the case of large molecules resort to the use of resonance scatteres, in order to increase the contrast from these scatterers. Such kind of scatterers exhibit a characteristic and most advantageous dependence of the scattering amplitude and its phase on the wavelength.

There exist at present three types of resonance scattering techniques which are potentially useful for solving the phase problem in the structure analysis of crystalline proteins:

a) X-ray resonance scattering (anomalous dispersion), employing X rays with wavelengths in the vicinity of absorption edges of specific heavy atoms inside the unit cell.

b) Neutron resonance scattering employing nuclear isotopes with neutron resonances in the range of thermal neutron energies.

Table 2.2. Comparison of structure factors

| | | Structure factor $\|F\|$ electron equivalents | $10^{-12}$ cm |
|---|---|---|---|
| X ray | free electron | 1 | 0.282 |
| | C atom | 6 | 1.692 |
| | Myoglobin $<\|F_P\|>$ | 235 | 66.27 |
| | Hg atom | 80 | 22.56 |
| anomalous dispersion of X rays | Fe; K-edge $\lambda_1 = 1.66$ Å; $\lambda_2 = 1.79$ Å | 3.5 | 0.987 |
| | La; L-edge $\lambda_1 = 1.93$ Å; $\lambda_2 = 2.28$ Å | 10.0 | 2.82 |
| | Hg; both energies above the L-edge $\lambda_1 = 0.78$ Å ; $\lambda_2 = 0.83$ Å | 5.5 | 1.55 |
| $\gamma$ rays | $^{57}$Fe (upper limit) | 520 | 146.64 |
| | $^{57}$Fe (lower limit) $\lambda = 0.86$ Å | 166 | 46.81 |
| neutrons | C atom | | 0.665 |
| | Myoglobin $<\|F_P\|>$ | | 25.6 |
| | $^{149}$Sm $\lambda_1 = 0.78$ Å; $\lambda_2 = 1.12$ Å | | 6.39 |

c) Gamma-ray resonance scattering, employing recoilless nuclear gamma transitions (Mössbauer effect) in isotopes with very low resonance energies.

The wavelengths employed in any of these techniques must be of the order of 1 Å, since one is aiming at atomic resolutions. All resonance methods have a distinct common advantage which is absent in the standard method of isomorphous replacement. The scattering power of the resonance reference scatterer is easily modified by changing the wavelength of the incident radiation and it can in principle be turned on and off. Thus it becomes possible to perform phase determinations on one and the same crystal and there is no problem of isomorphism. A phase determination, by consequence, can become correspondingly more precise. Examples for different reference scatterers are given in Table 2.2. A comparison is given there of the structure amplitudes of a C atom, a Hg atom and of the mean structure factor of a myoglobin crystal (M=17000 g/mol).

Structure determinations by means of resonant scattering techniques proceed in the same manner as has been outlined above for the case of the heavy-atom isomorphous replacement method. Intensity measurements are combined with experimental phase determinations. One usually employs powerful X-ray sources for the intensity measurements, while phase determinations are performed with one of the resonance techniques.

### 2.2.2 General Features of Phase Determinations by Means of Gamma-Resonance Scattering

Table 2.2 shows a dramatic increase in the resonance scattering amplitude in going from X-ray to neutron to gamma-ray resonant scattering. The intensities of the available radiation sources, by contrast, decrease dramatically as one goes from X rays to neutron to gamma rays. In the one extreme case of X-ray resonance scattering, one is therefore dealing with very small resonance scattering effects which can be studied with very intense radiation sources, while at the other extreme of gamma-ray resonance scattering very large scattering effects prevail, but only weak radiation sources are available. This situation is more quantitatively illustrated in Table 2.3 for the case of scattering from a myoglobin crystal. The gamma-resonance method, while being hampered by severe brightness problems of the available radiation sources, nevertheless appears suitable for solving the phase problem in cases where other methods can no longer be used, i.e., for very heavy proteins. We summarize in this context, the following number of advantages which distinguishes the gamma-resonance method from the X-ray anomalous dispersion technique:

a) A phase determination for a particular Bragg reflection necessitates a measurement of the scattered intensity for at least three different frequencies, one far-off resonance, the other two within the resonance regime. The gamma-resonance method makes it possible to perform these three measurements for a fixed orientation of the crystal. This way many corrections for varying experimental conditions can be avoided in the determination of a single phase.

Table 2.3. The "contrast" $\eta_c$ obtained from different reference scatterers. $\eta_c = (I_{P+R}-I_P)/I_P$. Two molecules and two reference scatterers are assumed to be in the unit cell. The ratio of the numbers of the different types of atoms in the molecule is taken to be equal to that for sperm whale myoglobin. The first column refers to isomorphous replacement with Hg, the second to anomalous dispersion of X rays at the L-edge of La, and the following columns refer to the resonance scattering of neutrons or of $\gamma$ rays, respectively. The intensity $I_0$ of the primary beam for X and $\gamma$ rays is discussed in Sect.2.3. The value of neutrons is taken from [2.22,23]

| M [g/mol] | $\eta_c$ values | | | | |
|---|---|---|---|---|---|
| | Hg | La | $^{149}$Sm | $^{57}$Fe upper limit | $^{57}$Fe lower limit |
| 17 816 | 0.48 | 0.059 | 0.356 | 3.09 | 0.99 |
| 100 000 | 0.20 | 0.025 | 0.150 | 1.30 | 0.42 |
| 250 000 | 0.13 | 0.016 | 0.095 | 0.82 | 0.26 |
| 500 000 | 0.09 | 0.011 | 0.067 | 0.58 | 0.19 |
| 1 000 000 | 0.06 | 0.008 | 0.047 | 0.41 | 0.13 |
| $I_0$ [$cm^{-2} \times s^{-1}$] | $2 \times 10^9$ | | $10^7$ | $6 \times 10^5$ | |

b) Frequency changes are most easily achieved in the Mössbauer effect by means of Doppler shift techniques. The frequencies are thereby easily changed to any point in the vicinity or far-off the resonance. Such frequency changes are much more difficult to achieve with X-ray sources, although the use of synchrotron radiation or of rotating anode X-ray tubes plated with different metals [2.24,25] may remedy this situation in the future.

c) The frequency changes necessary for the gamma-resonance method are extremely small, of the order $10^{-8}$ eV. By consequence there are no changes with frequency in the absorption or scattering of the nonresonant radiation, which substantially simplifies the data analysis.

d) Gamma-ray sources, apart from the well-known and readily accountable intensity diminuation due to finite half-lives, show in principle no long-time nonstatistical fluctuations in their intensities. Their intensity stability therefore largely exceeds that of X-ray generators.

e) The very large resonance scattering amplitudes typical for the gamma-resonance method yield the same accuracy in the phase determination for much smaller numbers of scattered photons than is the case with X-ray anomalous scattering. This obviously results in great advantages as far as radiation damage problems are concerned.

f) Iron occurs naturally in many active centers of proteins. Structure determinations on such systems permit measurements on molecules which have not been chemically manipulated, i.e., which exist in their native state. The implantation problem of the resonant nuclei then reduces to a mere isotope exchange problem, requiring an enrichment in the resonant isotope $^{57}$Fe at the expense of the naturally abundant $^{56}$Fe.

g) The gamma-resonance method employs a single extremely narrow and very well-defined resonance line, in contrast to the X-ray case, where the frequency distribution is given by a rather broad continuum.

### 2.2.3 Resonance Scattering of Gamma Radiation

We summarize in this section certain features of the resonance scattering of gamma radiation which are of relevance to a solution of the phase problem of the structure analysis of protein crystals. The 14.4-keV gamma radiation in $^{57}$Fe is resonantly scattered by the $^{57}$Fe nuclei in the protein single crystal, the structure of which one wishes to analyze. This energy corresponds to a wavelength of 0.86 Å. The natural linewidth of the 14.4 keV transition, which is observed in the Mössbauer effect, is only $4.7\times10^{-9}$ eV, corresponding to a half-life for the first excited state in $^{57}$Fe of $10^{-7}$ s.

We are primarily interested in the scattering amplitude $n(E,E_\ell)$ which describes the coherent resonance scattering of an incident photon with wave vector $\underline{k}_1$, polarization unit vector $\hat{\pi}_1$, and energy E into an outgoing photon with wave vector $\underline{k}_2$ and polarization unit vector $\hat{\pi}_2$ while the nucleus undergoes a transition between nuclear states with energy separation $E_\ell$. $\hat{\pi}$ refers to the electric field. We shall make the usual assumption, well justified in the Mössbauer effect, that we may factorize the relevant transition matrix elements into factors involving either the center of mass motion of the nuclei (lattice dynamical factor) only or the internal nucleonic motions (nuclear factor). We then obtain, assuming for the moment only one nucleus per elementary cell

$$n(E,E_\ell) = K \frac{\langle I_g M_g|H'^*(\underline{k}_2)|I_e M_e\rangle\langle I_e M_e|H'(\underline{k}_1)|I_g M_g\rangle}{E-E_\ell+i(\Gamma/2)} f(\underline{k}_1)f(\underline{k}_2) \quad . \tag{2.7}$$

The quantum numbers $I_g$, $I_e$, $M_g$, $M_e$ specify respectively the spin and magnetic quantum numbers of the ground (g) and excited (e) state of the nucleus, while $\Gamma$ is the total natural width of the nuclear excited state and K is a proportionality constant. The factor $f(\underline{k}_1)f(\underline{k}_2)$ is the Lamb-Mössbauer factor which will be discussed later on. The second matrix element in (2.7) describes the absorption (annihilation) of a photon $\underline{k}_1$, while the successive reemission (creation) process of a photon $\underline{k}_2$ is given by the first matrix element. The interaction Hamiltonian $H_i' = \sum_n \frac{e}{m} \underline{p}_n \underline{A}_i(\underline{k}_i)$ is linear in the vector potential $\underline{A}$ of the electromagnetic field interacting with nucleons of charge e, mass m and linear momentum $\underline{p}_n$. We shall use a plane wave representation for the vector potential, with $\underline{A}(\underline{k}_1) = A_1\hat{\pi}_1 \exp(i\underline{k}_1\underline{r})$ for the incoming and $\underline{A}(\underline{k}_2) = A_2\hat{\pi}_2^* \exp(-i\underline{k}_2\underline{r})$ for the outgoing wave. The exponent $\underline{k}_i\underline{r}$ is practical constant over the nuclear volume for the gamma-radiation wave vectors of interest here and we may consequently expand the exponential factor. Confining our present treatment to dipole transitions only allows us to limit this expansion to the first term. The absorption and emission matrix elements, respectively, then reduce to

$$M_1 = \left\langle I_e M_e \middle| \sum_n \frac{e}{m} \underline{p}_n A_1 \hat{\pi}_1 \middle| I_g M_g \right\rangle \quad , \tag{2.8a}$$

$$M_2 = \left\langle I_g M_g \middle| \sum_n \frac{e}{m} \underline{p}_n A_2 \hat{\pi}_2^* \middle| I_e M_e \right\rangle \quad . \tag{2.8b}$$

This expression simplifies further, if we employ the Wigner-Eckart theorem which allows one to pull out of the matrix element a factor which carries all the dependences on the magnetic quantum numbers. It is thereby necessary to transform the scalare product in (2.8) into the form of a spherical tensor, the components of which correspond to definite magnetic quantum numbers. We introduce for this purpose the spherical basis vector

$$\hat{e}_{\pm 1} = \mp \frac{1}{\sqrt{2}} (\hat{e}_x \pm i\hat{e}_y) \quad \text{and} \quad \hat{e}_0 = \hat{e}_z \quad . \tag{2.9}$$

The spherical components $a_Q^{(1)}$ of a vector $\underline{a}$ which transform under rotations like a spherical tensor of rank 1, i.e., the spherical harmonics $Y_M^{(1)}(\Theta,\phi)$, are then **given** by

$$a_Q^{(1)} = (\underline{a}\hat{e}_Q) \quad , \text{ where } \quad \underline{a} = \sum_{Q=-1}^{+1} (-1)^Q a_{-Q}\hat{e}_Q \quad . \tag{2.10}$$

The scalar product of two vectors $\underline{a}$ and $\underline{b}$ is consequently given by

$$(\underline{a}\underline{b}) = \sum_{Q=-1}^{+1} (-1)^Q a_Q b_{-Q} \quad . \tag{2.11}$$

Application of this formalism to (2.8) yields immediately

$$M_1 = \left\langle I_e M_e \middle| \sum_n \frac{e}{m} A_1 \sum_Q (-1)^Q (\underline{p}_n \hat{e}_Q)(\hat{\pi}_1 \hat{e}_{-Q}) \middle| I_g M_g \right\rangle \quad ,$$

$$M_2 = \left\langle I_g M_g \middle| \sum_n \frac{e}{m} A_2 \sum_{Q'} (-1)^{Q'} (\underline{p}_n \hat{e}_{Q'})(\hat{\pi}_2^* \hat{e}_{-Q'}) \middle| I_e M_e \right\rangle \quad .$$

We have now transformed to spherical tensors and may therefore apply the Wigner-Eckart theorem obtaining

$$M_1 = \sum_Q (-1)^Q \langle I_g 1 M_g {-Q} | I_e M_e \rangle \langle I_e \| H_1' \| I_g \rangle (\hat{\pi}_1 \hat{e}_{-Q}) \quad ,$$

$$M_2 = \sum_{Q'} (-1)^{Q'} \langle I_e 1 M_e {-Q'} | I_g M_g \rangle \langle I_g \| H_2'^* \| I_e \rangle (\hat{\pi}_2^* \hat{e}_{-Q'}) \quad . \tag{2.12}$$

The Clebsch-Gordan coefficients $\langle I_g 1 M_g - Q | I_e M_e \rangle$ and $\langle I_e 1 M_e - Q' | I_g M_g \rangle$ differ from zero only for $Q = M_g - M_e$, $Q' = M_e - M_g$, respectively, while the reduced matrix elements $\langle I_e \| H_1' \| I_g \rangle$ and $\langle I_g \| H_2'^* \| I_e \rangle$ are independent of magnetic quantum numbers. We then obtain for the products of the Hermitian conjugate emission and absorption matrix elements which appear in (2.7):

$$M_2 \cdot M_1 = \sum_Q (-1)^{Q+I_e-I_g} \frac{1}{2I_e+1} \langle I_g 1 M_g {-Q} | I_e M_e \rangle^2$$

$$\cdot \langle I_g \| H_2'^* \| I_e \rangle \langle I_e \| H_1' \| I_g \rangle (\hat{\pi}_2^* \hat{e}_Q)(\hat{\pi}_1 \hat{e}_{-Q}) \quad ,$$

where we have used

$$\langle I_e 1 M_e {-Q'} | I_g M_g \rangle = (-1)^{I_e - I_g - Q'} \sqrt{\frac{2I_g+1}{2I_e+1}} \langle I_g 1 M_g {-Q'} | I_e M_e \rangle$$

and $Q' = -Q$. We shall now insert $M_2 M_1$ into (2.7) while simultaneously incorporating into the proportionality constant K all factors which depend only on the spin quantum numbers $I_e$ and $I_g$:

$$n(E,E_\ell) = \tilde{K} \frac{\sum_Q (-1)^Q \langle I_g 1 M_g {-Q} | I_e M_e \rangle^2 (\hat{\pi}_2^* \hat{e}_Q)(\hat{\pi}_1 \hat{e}_{-Q})}{E - E_\ell + i(\Gamma/2)} f(\underline{k}_1) f(\underline{k}_2) \quad . \tag{2.13}$$

The constant $\tilde{K}$ may now be determined with the help of the optical theorem which relates the scattering amplitude in the forward direction to the familiar total mean cross section $\sigma_t$ (averaged over the isotopic composition) for the Mössbauer effect. We have

$$\sigma_t = \frac{4\pi}{k_1} \, \mathrm{Im}\left\{n_{coh}(E,E_\ell,\underline{k}_1=\underline{k}_2)\right\} \quad , \tag{2.14}$$

where we have adopted the usual sign convention for $\gamma$-ray amplitudes and where

$$\sigma_t = \frac{2\pi}{k_1^2} \frac{2I_e+1}{2I_g+1} \frac{\eta}{1+\alpha} \frac{\left(\frac{\Gamma}{2}\right)^2}{(E-E_\ell)^2-\left(\frac{\Gamma}{2}\right)^2} f(\underline{k}_1)^2 \quad . \tag{2.15}$$

The expression contains the internal conversion coefficient $\alpha$ and the relative isotopic abundance $\eta$. Equation (2.15) applies to a situation without any hyperfine splittings; the nuclear states are then completely degenerated. By consequence, we must use in (2.14) the coherent scattering amplitude for a gamma transition without hyperfine splitting. This amplitude follows immediately by summing the amplitude for an individual hyperfine component, $n(E,E_\ell)$, over all intermediate (i.e., excited) states

$$n_{coh}(E,E_\ell) = \sum_{M_e} n(E,E_\ell) \tag{2.16}$$

In performing the summation for the case of forward scattering we have to evaluate the expression

$$\begin{aligned}
\sum_{M_e} \sum_{Q} (-1)^Q < I_g 1 M_g -Q | I_e M_e >^2 & \quad (\hat{\pi}_1^* \hat{e}_Q)(\hat{\pi}_1 \hat{e}_{-Q}) \\
&= \sum_{Q} \sum_{M_e} (-1)^Q \frac{2I_e+1}{3} < I_g 1 M_e - M_g | I_e -Q >^2 (\hat{\pi}_1^* \hat{e}_Q)(\hat{\pi}_1 \hat{e}_{-Q}) \\
&= \sum_{Q} (-1)^Q \frac{2I_e+1}{3} (\hat{\pi}_1^* \hat{e}_Q)(\hat{\pi}_1 \hat{e}_{-Q}) = \frac{2I_e+1}{3} \quad ,
\end{aligned}$$

where we have used the fact that in the forward direction we have $\underline{k}_1 = \underline{k}_2$ and $\hat{\pi}_1 = \hat{\pi}_2$; we thus obtain

$$n_{coh}(E,E_\ell,\underline{k}_1=\underline{k}_2) = \tilde{K} \left(\frac{2I_e+1}{3}\right) f_1^2(\underline{k}_1) \cdot \frac{(E-E_\ell)-i\frac{\Gamma}{2}}{(E-E_\ell)^2+\left(\frac{\Gamma}{2}\right)^2} \quad . \tag{2.17}$$

Introducing (2.15,17) into (2.14) finally yields

$$\tilde{K} = -\frac{3}{2k_1} \frac{\eta}{1+\alpha} \frac{1}{2I_g+1} \frac{\Gamma}{2} \quad . \tag{2.18}$$

The coherent scattering amplitude follows in the general case ($\underline{k}_1 \neq \underline{k}_2$) from (2.13) with $\tilde{K}$ given by (2.18). This case applies to the situation where all hyperfine components coincide energetically (unsplit line). Extending this case to a situation with more than one resonant nucleus per unit cell we obtain

$$N(E,E_\ell) = \sum_\rho n(E,E_\ell)\, \exp[i(\underline{k}_2-\underline{k}_1)\underline{r}_\rho] \quad , \qquad (2.19)$$

where $\underline{r}_\rho$ specifies the position of the resonant nucleus within the unit cell. The other limiting case is represented by the situation where one deals with a completely resolved magnetic hyperfine splitting. In this case one uses only one value for $M_g$ and Q in (2.3) specifying the particular hyperfine component.

If one employs for phase determination the gamma-resonance scattering technique instead of the isomorphous replacement technique then $F_R$ in (2.6) has to be replaced by $N(E,E_\ell)$ as given by (2.19), and one must take into account the energy distribution of the incident radiation.

It should be noted that the coherent nature of the scattering process is contained in the fact that the initial and final states of the scattering process have been chosen identical, as is clearly indicated in (2.7). It should also be noted that the polarization vectors $\hat{\pi}_1$ and $\hat{\pi}_2$ in (2.19) are unit vectors in the direction of the electric field in the case of electric dipole transitions, while in the case of magnetic dipole transitions, which are of prime interest in this paper, the replacement $\hat{\pi}_i \Rightarrow (\hat{\pi}_i \times \underline{k}_i)/\underline{k}_i$ with $i = 1,2$ must be made.

The symmetry of the local site of the resonant nucleus inside a protein molecule will usually be rather low and the presence of quadrupole hyperfine splittings will be typical. In the case of an axially symmetric electric field gradient one will naturally choose the gradient axis as the most simple axis of quantization. In the general case typical for protein crystals, all three components $v_{xx}$, $v_{yy}$, and $v_{zz}$ of the electric-field-gradient tensor are different, though they obey the Laplace equation $v_{xx}+v_{yy}+v_{zz}=0$. In this general case the scattering amplitude for $^{57}Fe$ is obtained as follows. The orthonormalized functions for the two excited quadrupole-split levels of the spin $I_e = 3/2$ state are given by

$$|\psi^+\rangle = \cos\mu\left[\left|\tfrac{3}{2},\tfrac{3}{2}\right\rangle+\left|\tfrac{3}{2}-\tfrac{3}{2}\right\rangle\right] + \sin\mu\left[\left|\tfrac{3}{2}-\tfrac{1}{2}\right\rangle+\left|\tfrac{3}{2},\tfrac{1}{2}\right\rangle\right] \quad ,$$

$$|\psi^-\rangle = -\sin\mu\left[\left|\tfrac{3}{2},\tfrac{3}{2}\right\rangle+\left|\tfrac{3}{2}-\tfrac{3}{2}\right\rangle\right] + \cos\mu\left[\left|\tfrac{3}{2}-\tfrac{1}{2}\right\rangle+\left|\tfrac{3}{2},\tfrac{1}{2}\right\rangle\right] \quad ,$$

where

$$\mathrm{tg}\,2\mu = (v_{xx}-v_{yy})\,/\,(\sqrt{3}\, v_{zz}) \quad .$$

The matrix elements in (2.8) are now given by

$$M_1^\pm = \left\langle I_e\psi^\pm \middle| \sum_n \frac{e}{m}\, \underline{p}_n A_1 \hat{\pi}_1 \middle| I_g M_g \right\rangle$$

$$M_2^\pm = \left\langle I_g M_g \middle| \sum_n \frac{e}{m}\, \underline{p}_n A_2 \hat{\pi}_2^* \middle| I_e\psi^\pm \right\rangle \quad .$$

Care has to be exercised in evaluating the product $M_2 \cdot M_1$ since the condition $Q' = -Q$ no longer holds. Detailed expressions will be given elsewhere [2.26]. As a result one obtains the scattering amplitudes $n^+(E,E_+)$ and $n^-(E,E_-)$ for the two quadrupole lines. Different amplitudes of this kind have to be evaluated in case the molecule contains more than one resonant nucleus in sites differing in symmetry. Even in the case of myoglobin where the molecule contains only one iron, $n_1^{\pm}(E,E_{\pm})$ becomes different for the two molecules of the unit cell because of the different orientations of the coordinate system of the electric field gradient with respect to $\hat{\pi}_2$ and $\hat{\pi}_1$.

The treatment given here may be compared with other derivations of the nuclear resonance scattering amplitude [2.27-29]. It also should be noted that the evaluation of electric field gradients may lead to ambiguities. This situation arises in the case of several identical nuclei within the unit cell which transform into each other by symmetry operations [2.30-32].

The gamma radiation resonantly scattered with the coherent amplitude $F_R = N(E,E_\ell)$ will interfere with the gamma radiation which is nonresonantly scattered by the electronic charge distribution (Rayleigh scattering) in the unit cell of the crystal. The scattering for the latter process is given by

$$F_P(E) = -\sum_\rho (f_{0\rho} + f'_{0\rho} + if''_{0\rho}) r_0 f(\underline{k}_2 - \underline{k}_1)(\hat{\pi}_2^* \cdot \hat{\pi}_1) \exp[i(\underline{k}_2 - \underline{k}_1)\underline{r}_\rho] \quad , \tag{2.20}$$

where $f_{0\rho}$ is the atomic form factor of the $\rho^{th}$ atom of the unit cell, $f'_{0\rho}$ and $f''_{0\rho}$ are the usual correction terms for anomalous dispersion effects, $r_0$ is the classical electron radius and $f(\underline{k}_2 - \underline{k}_1)$ is the Debye-Waller factor, while all other symbols have been introduced previously.

We conclude this section with a few remarks about the lattice-dynamical parameters $f(\underline{k})$ which enter in the expressions for the scattering amplitudes described by (2.19, 20). The factor

$$f_{LM} = f(\underline{k}_1) \cdot f(\underline{k}_2) \quad , \tag{2.21}$$

which enters in the nuclear matrix elements, is the Lamb-Mössbauer factor. In the case of an isotropic motion of the scattering nucleus, $f_{LM}$ becomes

$$f_{LM} = f(\underline{k}_1)^2 = f(\underline{k}_2)^2 = f' \quad ,$$

the notation conventionally used in nuclear gamma-resonance absorption spectroscopy. The analogous expression

$$f_{DW} = f(\underline{k}_2 - \underline{k}_1)$$

is the Debye-Waller factor. Both quantities $f_{LM}$ and $f_{DW}$ directly give the percentage fraction of the scattering processes which occur elastically. The distinction between both factors arises from the fact that gamma-resonance scattering (Mössbauer scattering) is a resonance process, which therefore proceeds via an intermediate state characterized by a finite lifetime, while the electronic scattering (Rayleigh scattering) is a direct first-order process which proceeds instantaneously. Rayleigh scattering, by consequence, certainly occurs in times short compared to typical characteristic dynamic time constants of the scattering system. In other words, the transfer of the recoil momentum $\hbar(\underline{k}_2-\underline{k}_1)$ occurs in a single step and the Debye-Waller factor becomes a function of $(\underline{k}_2-\underline{k}_1)$. Gamma-resonance scattering, by contrast, is associated with lifetimes in the intermediate (excited) state of the nucleus which are long compared to typical vibration times in the lattice which are of order $10^{-13}$ s. The gamma-scattering process therefore exhibits a separate dependence on the momentum transfer $\hbar\underline{k}_1$ and $\hbar\underline{k}_2$, and $f_{LM}$ becomes dependent on the product of two factors $f(\underline{k}_1)f(\underline{k}_2)$, one each for the excitation and the deexcitation process [2.33].

## 2.3 Experimental Details

### 2.3.1 The Gamma-Ray Source

The use of the $^{57}$Fe nucleus as a reference scatterer for the phase determination requires replacement of the X-ray tube by a $^{57}$Co gamma-ray source emitting the 14.4 keV radiation of $^{57}$Fe. Figure 2.3 shows the $\vartheta/2\vartheta$ scan of the $60\bar{3}$ reflection of a myoglobin crystal performed with a $^{57}$Co source [2.34]. The matrix material was a 12 μ foil of Rh. 200 mCi of $^{57}$Co were deposited on an area of $2\times2.5\,\mathrm{mm}^2$. At the time of the experiment the activity had decreased to about 100 mCi. The source surface was inclined by 30° with respect to the line connecting the source and the crystal, so that an effective area of $1\times2.5\,\mathrm{mm}^2$ was seen by the crystal. The counting rate of 1 c/min in the peak at $\vartheta = 2.77°$ has to be compared with the corresponding value of $0.63\times10^5$ c/min obtained with the $K_\alpha$-line of a conventional Mo X-ray tube (50 KV, 12 mA). Thus a reduction of the counting rate by a factor of $1.6\times10^{-5}$ occured when the X-ray tube was exchanged by the $^{57}$Co source. The question arises, how the inten-

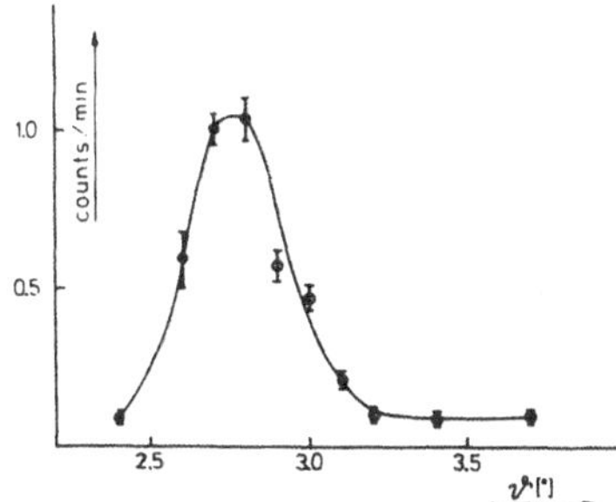


Fig. 2.3. $\vartheta/2\vartheta$ scan of the $60\bar{3}$-reflection of a myoglobin crystal performed with a $^{57}$Co source

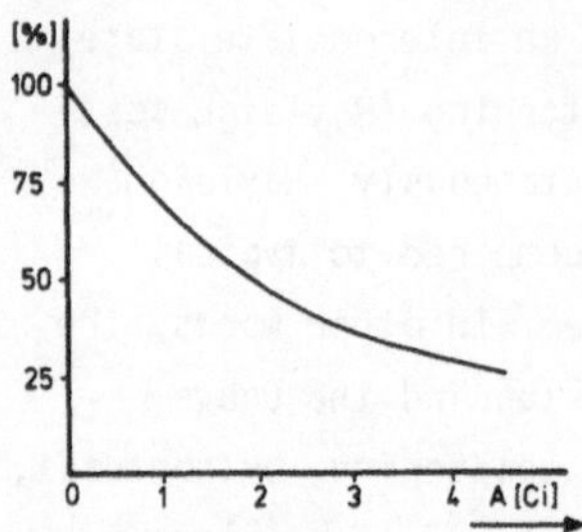


Fig. 2.4. Reduction of the intensity of a $^{57}$Co source as a function of its activity. x-axis: $^{57}$Co activity deposited on 1 mm$^2$. y-axis: percentage of the activity which can be used for the experiment after taking into account the electronic self-absorption in the Co atoms

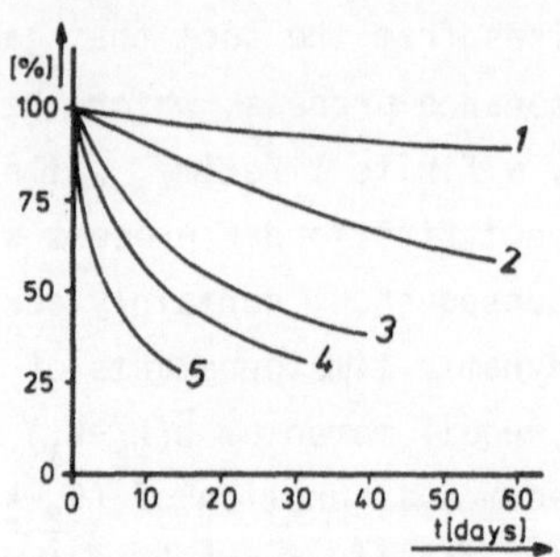


Fig. 2.5. Reduction of the emitted 14.4 keV radiation of a $^{57}$Co source by nuclear resonance absorption as a function of the time t (days) after the preparation. 1) No nuclear resonance absorption; 2) 100 mCi/mm$^2$; 3) 400 mCi/mm$^2$; 4) 750 mCi/mm$^2$; 5) 2500 mCi/mm$^2$

sity of such a source can be increased. Unfortunately, there are principal limiting factors according to area and thickness of such a source.

As a rule the area of the source should be the same as the area of the crystal which is illuminated by the beam. Otherwise the radiation of the edges of the source arrives at the crystal at an angle where the Bragg condition is no longer fulfilled. For convenience we consider, in the following, protein crystals with a volume of 1 mm$^3$ and a surface of 1 mm$^2$.

The reasonable thickness of a source is governed by considerations of electronic self-absorption. If we increase the amount of $^{57}$Co per 1 mm$^2$ we increase at the same time the electronic absorption of the 14.4-keV radiation within the source material. Figure 2.4 gives this relation. If we deposit 2.5 Ci $^{57}$Co on 1 mm$^2$ area, 14.4-keV radiation corresponding to an effective activity of only 1 Ci becomes available for the experiment. The difference is consumed by self-absorption. The situation is complicated by the continuous generation of $^{57}$Fe as a consequence of the $^{57}$Co decay. Thus the nuclear resonance self-absorption builds up as a function of the time after the preparation of the source as shown in Fig.2.5. By consequence, the $^{57}$Co source of 2.5 Ci/mm$^2$ activity emits after 5 days only one-half of the initial intensity. At the same time the linewidth of the emitted Lorentzian increases by a factor of two. It should be noted that the facts presented in Fig.2.5 apply to sources with natural linewidth. Line-broadening, of course, reduces the self-absorption. As a consequence, the $^{57}$Co sources are limited from a practical point of view to an effective activity of 1 Ci/mm$^2$. An experiment requires then two sources, one which is employed 5 days in the experiment and a second one which during this time is purified by extraction of the $^{57}$Fe contents.

A careful comparison of the primary intensity of an X-ray tube and an optimal $^{57}$Co source was performed. For this purpose the integral reflectivities were determined for beryllium acetate, sodium chloride and myoglobin crystals at several reflections employing the $K_{\alpha}$-line of a Mo tube. A voltage of 50 kV and a current of 12 mA were used. From these experiments we obtained a primary X-ray intensity of $2\times10^{9}$ $cm^{-2}$ $s^{-1}$ at the crystal, which had a distance of 37.5 cm from the X-ray tube. Several $^{57}$Co sources were used to determine the integral reflectivity of the $60\bar{3}$ reflection for a number of myoglobin crystals. The distance between the source and the crystal was 20 cm. In the average the scattered intensity was a factor of $3\times10^{-4}$ weaker for the gamma-ray sources normalizing to a source of an effective activity of 1 Ci/$mm^2$. Thus we may expect a primary intensity of $6\times10^{5}$ $cm^{-2}$ $s^{-1}$ for the optimal $^{57}$Co source. The values discussed above were used for Table 2.3.

The $^{57}$Co can be produced in a cyclotron by the bombardment of $^{58}$Ni with 20 MeV protons. In 100 h one obtains 1 mC $^{57}$Co per 1 $\mu$A proton current. The isotope $^{57}$Co must be radiochemically extracted.

For the scattering experiment one strives for a source with a large Lamb-Mössbauer factor $f_s$ emitting only one Lorentzian with the natural linewidth. Therefore, the $^{57}$Co atoms should be in a lattice with cubic symmetry in order to avoid quadrupole splittings. In the case of an optimal efficient $^{57}$Co source the ferromagnetic or antiferromagnetic coupling of the Co atoms cannot be prevented by dilution in a nonmagnetic matrix. Such a dilution would cause an intolerably high self-absorption in the source. One therefore has to use Co compounds at temperatures above the Curie or the Neel point. CoO could, in principle, fulfill all requirements for such a compound but the production of the stoichiometric cubic phase is by no means easy to reproduce [2.35-38]. An alternative is $CoSb_3$ [2.39], although the antimony content increases the self-absorption in the source.

### 2.3.2 The Lamb-Mössbauer Factor of a Protein Crystal

In Sect.2.2.3 the scattering amplitude of an $^{57}$Fe nucleus was derived in detail. It is proportional to the Lamb-Mössbauer factor f' as one sees from (2.19). An experimental determination yielded $f'=0.03$ [2.40,41] for myoglobin crystals at room temperature. This reduces $n(E,E_{\ell})$ of the $^{57}$Fe nucleus from 520 electron equivalents to about 16 which is too small for phase determination.

If one goes down to 4.2 K, f' rises to 0.8. Unfortunately, the cooling of protein crystals without damage is not a trivial problem, since about half of the volume of the crystal consists of water or buffer. During the water-to-ice phase transition, this water expands and damages the crystal. A method for freezing sperm whale myoglobin crystals without damage is described in [2.42]. The crystals are brought into isopentane which does not penetrate into the interior. Then a hydrostatic pressure of 2500 atm is applied, and the crystals are cooled to liquid $N_2$ temperature. Under these conditions a high-pressure ice phase is established which has a smaller volume than water. Moreover, the volume change during this phase transition is smaller than at 1 atm. A high-pressure ice phase at 77 K stays metastable for years, therefore the structure investigation now can be performed at 1 atm at low temperatures. As shown by precession photographs, the crystals are not distorted by this freezing procedure. A difference Fourier analysis with room-temperature data has shown that the structure of the myoglobin molecule remains practically unchanged. For relatively small crystals another method of freezing gave good results. Such crystals were frozen very rapidly by putting them into liquid propane as described in [2.32].

As a consequence of the low f' value at room temperature, all phase determinations using the nuclear resonance scattering have to be performed at low temperatures. One may combine these phases together with intensities measured at the same temperature and then obtain the low-temperature structure of the protein which supposedly is very similar to the room-temperature structure. Another possibility is the combination of low-temperature phases with room-temperature intensities. Since the experimental phases are used only in the beginning of the refinement procedure [2.43,44] one thereby obtains the room-temperature structure of the protein.

It should be mentioned, that the study of the temperature dependence of the Lamb-Mössbauer factor can give interesting information on the dynamic properties of the protein. According to recent investigations [2.45], a protein may exist in several conformational substates having practically the same energy. At room temperature these substates are in thermal equilibrium with permanent fluctuation between them. These fluctuations may be responsible for the large $\langle x^2 \rangle$ value which has been observed for $^{57}$Fe in myoglobin crystals at room temperature [2.40,41]. At low temperatures the fluctuation processes are frozen, and one observes again the same small $\langle x^2 \rangle$ values as are typical in inorganic or organic-iron compounds.

### 2.3.3 Phase Determination on a Myoglobin Crystal

In order to check the phase determination procedure by nuclear resonance scattering, an experiment on a CO-liganded sperm whale myoglobin crystal has been performed [2.34]. Figure 2.6 shows the experimental setup. The frozen myoglobin crystal was first oriented on a liquid nitrogen cryostat mounted on a 3-circle diffractometer ($\vartheta$, $2\vartheta$ and $\varphi$ circle). The orientation was controlled with the help of a Mo X-ray tube. The experiment was performed with a source as described in the beginning of Sect.2.3.1 and mounted on an electromechanical driving system. The source was carefully shielded by lead (5 cm in the forward direction) to reduce the background coming from the high-energy gamma radiation of the $^{57}$Co. The gamma quanta scattered into the $60\bar{3}$ reflection ($\vartheta=2.77^\circ$) were counted by a Si(Li)-solid state counter which was connected to the $2\vartheta$ circle of the diffractometer.

Figure 2.7 shows the counting rate for this reflection as a function of the Doppler velocity of the source. The experimental data have been least-squares fitted with the function:

$$R_c = C\left\{\int_{-\infty}^{+\infty} I_s(E,v)A(E)\left[|F_P|^2+2|F_P||F_R|\cos(\phi_P-\phi_R)+|F_R|^2\right]dE+|F_P|^2(1-f_s)\right\} \quad , \quad (2.22)$$

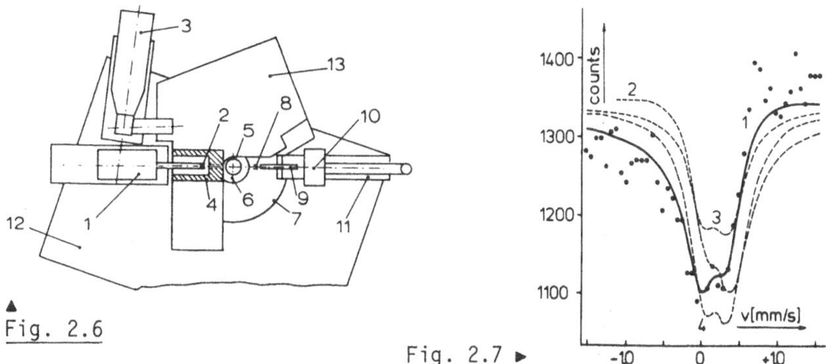


Fig. 2.6

Fig. 2.7 ▸

Fig. 2.6. Experimental setup for a phase determination. 1) Electromechanical driving system; 2) gamma-ray source; 3) X-ray tube which can be shifted to the position of the electromechanical driving system; 4) lead screening; 5) $N_2$-cryostat mounted on a cross-sliding carriage; 6) $\varphi$-circle of the diffractometer; 7) $\vartheta$ and $2\vartheta$ circle of the diffractometer; 8) beam stop; 9) collimator for the counter; 10) solid-state counter; 11) connection of the counter with the $2\vartheta$ circle of the diffractometer; 12) level 1 base plate; 13) level 2 base plate

Fig. 2.7. Intensity of 14.4 keV radiation scattered into the $60\bar{3}$ reflection of a myoglobin crystal as a function of the Doppler velocity of the source. 1) Least-squares fit yielding $\phi_P = -1.2$; 2) simulation with $\phi_P = 180^\circ$; 3) $\phi_P = 90^\circ$, and 4) $\phi_P = 270^\circ$

where C is a normalization constant, $|F_P|$ and $|F_R|$ are the absolute values of the structure factors, and $\phi_P$ and $\phi_R$ the phases of the protein and the $^{57}$Fe nuclei, respectively. $I_s(E,v)$ is the energy distribution of the source as a function of the Doppler velocity v, $f_s$ is the Lamb-Mössbauer factor of the source, and A(E) an energy-dependent absorption correction.

Equation (2.22) is essentially equivalent to (2.6), but it takes into account some corrections which are necessary in practice. Since the $^{57}$Co source emits a Lorentzian, the radiation scattered by the crystal is a convolution of the energy distributions of the source, $I_s(E,v)$, and of the scatterer $F_R = N(E,E_\ell)$. Furthermore, an energy-dependent absorption A(E) has to be included in the integral, since the complex nuclear scattering amplitude of the $^{57}$Fe is correlated by the optical theorem to an absorption cross section. Although the path length of the gamma beam through the crystals is different for each scattering volume element, a homogeneous average thickness of about 0.4 mm was assumed in the least-squares fit.

The reference scattering amplitude $|F_R|$ of the iron nuclei in the unit cell was also modified with respect to (2.13). Since the iron atoms in the different molecules do not have an absolute identical arrangement of their nearest neighbors, the resonance energies are also slightly different. As a result, the effective linewidth $\Gamma_a$ of the crystal exceeds the natural linewidth $\Gamma$, resulting in a decrease of the scattering amplitude by a factor $\Gamma/\Gamma_a$ which was taken into account in the evaluations. In myoglobin the iron is not in a cubic environment. Quadrupole splitting of the resonance energies therefore occurs. As discussed in Sect.2.2.3, a knowledge of the electric-field-gradient tensor is necessary for the calculation of the polarization factors of the two lines. Since this information was not available the different weight of the two resonance lines were taken from an absorption experiment on the crystal performed in the same orientation as in the scattering experiment. The term $|F_P|^2(1-f_s)$ takes into account the scattering of the radiation emitted inelastically from the source.

The function $R_c$ in (2.22) was least-squares fitted to the experimental data. A variation of the parameters C and $\phi_P$ was allowed yielding $\phi_P = -1^\circ$ (solid line in Fig.2.7). From the structure of myoglobin one calculates the value $\phi_P(60\bar{3}) = 0$, which is in excellent agreement. The dashed lines in Fig.2.7 were calculated with the phases $\phi_P = 180^\circ$ (curve 2), $\phi_P = 90^\circ$ (curve 3) and $\phi_P = 270^\circ$ (curve 4). All other parameters were kept constant as in curve 1. No agreement exists between these curves and the experimental data. We estimate from these simulations that the accuracy of the present phase determination is at least $\pm 45^\circ$, which is certainly good enough for a structure determination.

## 2.4 Future Prospects

At present no structure determination has yet been performed by gamma-resonance scattering. This is due to the large technical difficulties which arise from the low intensity of the available sources. In this section we wish to discuss a concept which may result in a reasonable time for a structure determination of a protein with a relative high molecular weight.

### 2.4.1 The Investigation of Bacterial Catalase

The power of the use of nuclear resonance scattering should be demonstrated by solving the structure of a molecule with a weight larger than 100000 g/mol. Provided the protein contains iron which can readily be exchanged by the isotope $^{57}Fe$, it should then be possible to determine the structure of a native protein. In order to circumvent a difficult isotope exchange by chemical means, one can grow bacteria or yeast in a medium containing only the isotope $^{57}Fe$. The proteins of interest are then prepared from these enriched materials. We employed this procedure with micrococcus luteus cells. These bacteria produce at optimal conditions up to 1 wt-% of the enzyme "catalase".

Catalase is a heme protein with a molecular weight of M = 240000 g/mol and contains 4 subunits, each of which has one iron atom in the active center. It catalyzes the decomposition of hydrogen peroxide into oxygen and water. Although bovine liver catalase had been crystallized many years ago [2.46-52], no structure analysis has been reported.

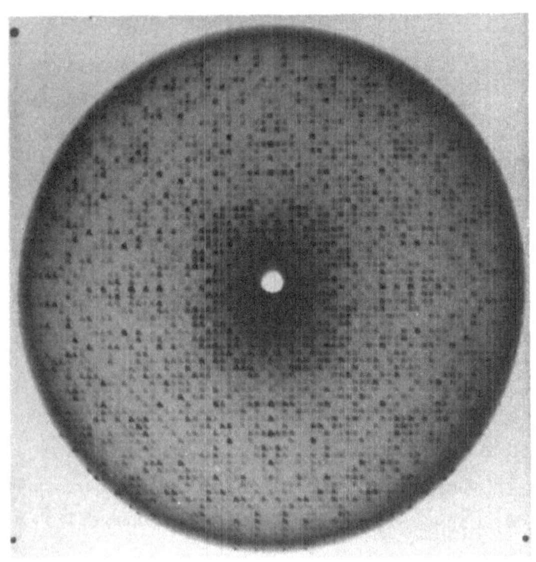

Fig. 2.8. Precession photograph of the (hk0)-plane of a catalase crystal. Cu $K_\alpha$-radiation filtered by a graphite monochromator was used. Note the four-fold symmetry axis

Catalase from micrococcus luteus was purified and crystallized yielding crystals with about 1 mm linear dimension [2.53]. Figure 2.8 shows the precession photograph of the (hk0) plane of such a crystal. The space group of the crystal is $P4_22_12$ [2.53]. The dimensions of the unit cell are a = b = 106.6 Å and c = 106.3 Å. The unit cell is composed of 8 asymmetric units which transform into each other by symmetry operations. It contains only two molecules and the four subunits of the catalase therefore are identical. The small number of molecules in the unit cell make the crystals very suitable for a structure determination. It should be stressed that bacterial catalase behaves quite differently than mammalian catalase since it contains no cysteine residue [2.54].

Let us give a rough estimate for the counting time of a phase determination on a reflection of the catalase crystal. In order to obtain an average structure factor we assume that catalase has the same ratio of C:N:O atoms as myoglobin, thus obtaining $\langle|F_P|\rangle = 1219$ electron equivalents. Since the unit cell contains eight $^{57}$Fe nuclei, we obtain, with the numbers of Table 2.2, a contrast $\eta_c = 0.54$. In order to get an estimate of the counting rate we make an extrapolation of our experimental values obtained from myoglobin crystals. The angular dependence of the counting rate for unpolarized radiation is given by $(1+\cos^2 2\vartheta)/(2\sin 2\vartheta)$. For an estimate we use $\vartheta = 8.8^\circ$ corresponding to a resolution of 2.8 Å. Assuming a $^{57}$Co source of 2.5 Ci/mm$^2$ activity and a catalase crystal with a volume of 1.4 mm$^3$, we estimate a counting rate of 0.1 c/min. Let us add a typical background of 0.1 c/min, we obtain 0.2 c/min. If we claim that the relative statistical error of the measurement is 0.06, we have to collect 278 quanta corresponding to a counting time of about one day. This time has to be multiplied by a factor of eight if we use for the phase determination eight different Doppler velocities of the source.

If we want to solve the structure with a resolution of 2.8 Å, the required number of reflections is about 230000. Because of symmetry relations it is necessary to measure only 1/8 of them. Nevertheless, it is obvious that the counting time becomes hopeless if one would try to measure the phase of the various reflections in succession. An experimental setup which allows measurement of a large number of reflections at the same time is apparently necessary. Such a device should also have a sufficiently good energy resolution in order to obtain a good signal-to-background ratio.

### 2.4.2 Two-Dimensional Position-Sensitive Proportional Counters

Figure 2.9 shows the reciprocal lattice of bacterial catalase with the well-known Ewald circle. Each reciprocal lattice point intersecting the circle originates a reflection. If the linear divergence of the beam is $\varepsilon$ degrees one obtains two limiting circles, or in the three-dimensional case, two spheres. All reciprocal lattice points between these spheres are excited at the same time, simultaneously yielding thousands of reflections. Their simultaneous measurement requires a new type of detector.

A photographic film, in principle, allows such a registration. But unfortunately, the peak-to-background ratio of a film is always very poor for low counting rates. This has two reasons. First of all, a film has a large thermal background which increases with the storage time. On the other hand, it is practically impossible to

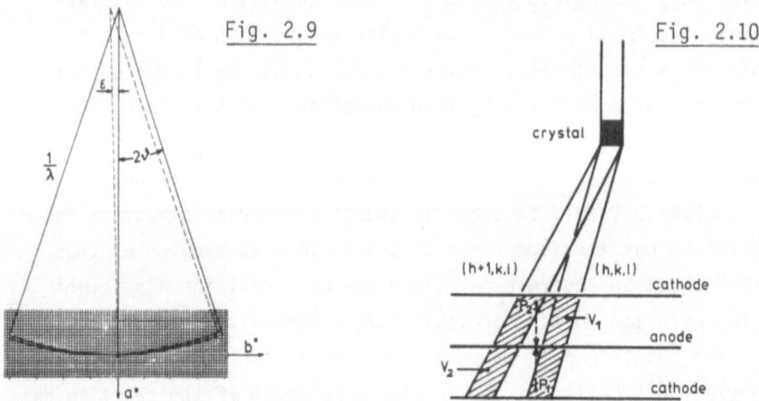


Fig. 2.9. Reciprocal lattice of bacterial catalase with Ewald circle. The drawing is in scale. All reciprocal lattice points between the two limiting circles originate simultaneous reflections, if the incoming beam has a divergence $\varepsilon$. The reflection (0380) corresponds to a spatial resolution of 2.8 Å

Fig. 2.10. Gamma quanta of two well separated reflections (hkℓ) and (h+1kℓ) enter a position-sensitive counter and are absorbed in the volumes $V_1$ and $V_2$, respectively. Electrons produced by the absorption process at $P_1$ and $P_2$, respectively, drift to the same position at the anode wire and cannot be attributed either to the reflection (hkℓ) or to (h+1kℓ)

count single events. Only the blackening due to a sufficiently large number of quanta falling on a small area can be converted into a number by photometric techniques.

The development of an area-sensitive solid-state counter was quite successful, but the number of reflections which can be registered by such a device is rather limited [2.55].

A position-sensitive proportional counter cannot easily be used for the simultaneous measurement of many Bragg reflections due to the inherent parallax problem. Figure 2.10 describes this problem. Two reflections which are well separated at the entrance of the counter cannot be resolved within the volume of a counter of sufficient thickness due to the overlap of the secondary electrons.

A very promising detector was conceived by CHARPAK et al. [2.56,57]. It should finally be available for the structure determination of bacterial catalase. In this detector the absorption of the gamma quanta and their registration is separated. The absorption volume is enclosed between the segments of two concentric spheres succeeded by a two-dimensional position-sensitive area detector. The crystal is located in the center of the two spheres. The quanta of a high-order reflection are scattered through the angle $2\vartheta$ and are absorbed in the volume between the segments of two spheres. A path length of 10 cm Xe is available for the absorption process thus yielding nearly 100% efficiency. A voltage of about 10 kV is applied enforcing a radial drift of the electrons which are generated by the absorption of the gamma rays. Therefore, the size of the reflection is not broadened by the absorption process (Fig.2.11).

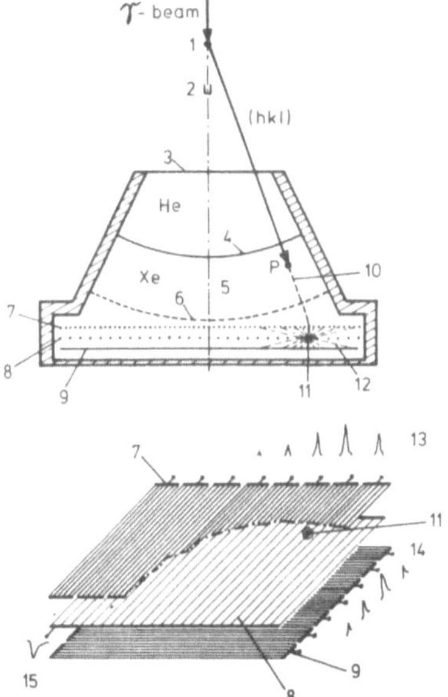


Fig. 2.11. Position-sensitive area detector with spherical drift chamber. 1) Crystal; 2) beam stop; 3) entrance window (mylar); 4) thin spherical Al-window; 5) Xe-filled absorption chamber; 6) spherical grid; 7) first cathode plane (6 wires are connected); 8) anode plane; 9) second cathode plane (6 wires are connected); 10) path of the electrons produced at P; 11) electron avalanche; 12) path of the ions drifting to the cathode planes; 13) pulse distribution at strips of the first cathode plane yielding the x-coordinate; 14) pulse distribution at strips of the second cathode plane yielding the y-coordinate; 15) pulse at the anode plane yielding the energy of the absorbed $\gamma$ quantum

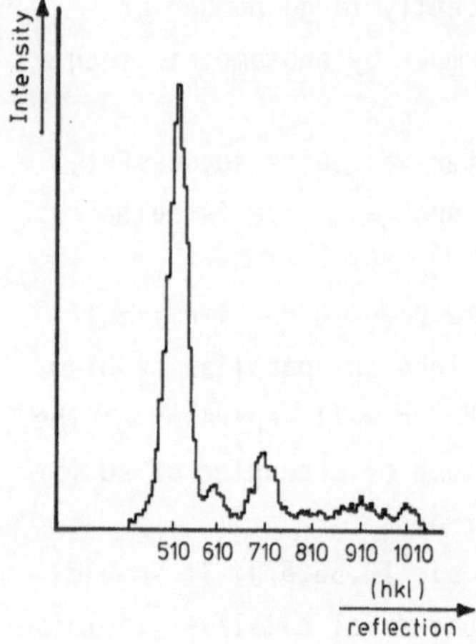


Fig. 2.12. Intensity (arbitrary units) of a linear section through the counter, covering the (h10) reflections of myoglobin. A raster of 0.78 × 0.78 mm was used during the data collection. The distance of two neighboring reflections corresponds to a distance of about 11.3 mm at the anode plane of the counter

The electrons are drawn by an appropriate voltage out of the drift chamber into a two-dimensional position-sensitive proportional counter. The prototype just under investigation has a size of 500 × 500 mm. It consists of three planes of counting wires. The anode plane contains 240 wires with distances of 2 mm and with 20 μm diameter. The cathode planes on either side are 6 mm apart. They contain 500 wires each with a diameter of 100 μm and a spacing of 1 mm. It is essential that the wires of the two cathode planes are fixed perpendicular to each other (compare Fig.2.11). An electron coming from the drift chamber travels to the nearest point on the anode plane. An electron avalanche builds in a region of about 40 μm linear dimension around the anode wire. The electrons of this avalanche are collected by one anode wire and are used for energy resolution. The registration of their xy coordinates, i.e., their position, is performed at the cathode planes in the following manner. The ions of the avalanche travel to the two cathode planes inducing during this process a current in a number of wires. This current is a function of the distance of these wires from the origin of the avalanche. The currents of the wires are amplified and used in an analog unit to calculate the center of gravity of the currents. This center of gravity determines the coordinates of the avalanche. The x-coordinate is taken from one cathode plane and the y-coordinate from the other. First measurements have shown that such a counter is well suitable for the structure determination of large proteins. Figure 2.12 gives the results of a preliminary scattering experiment with myoglobin crystals. The intensity of a linear section through the counter is shown. The measurement was performed with Mo $K_\alpha$ radiation.

We now go back to the estimate of the counting time for the phase determination of bacterial catalase. Simple geometrical considerations show that a divergence $\varepsilon$ of the primary beam of $1.5^\circ$ can be easily tolerated without getting substantial overlap of the reflections. An area-sensitive proportional counter with an area of 650 × 650 mm can collect all reflections excited simultaneously with this divergence.

Since one has to turn the crystal only through an angle of 90° to get all independent reflections, one has to perform 60 measurements. Using the estimates of Sect.2.4.1 we arrive at a counting time of 480 days for a structure with 2.8 A resolution. This is still rather long, but the time can again be reduced, for example, by working with one source which illuminates four crystals at the same time. One then needs four detectors, but the total time is reduced to 120 days. This is quite reasonable for a structure determination.

The considerations of Sect.2.4 show that the use of $^{57}Fe$ as a reference scatterer for phase determinations may become possible in the future. Nevertheless, a further development of the technique is necessary before the method can be turned toward practical uses.

*Acknowledgment.* The work is sponsored by the "Bundesministerium für Forschung und Technologie".

## References


2.1 D.W. Green, V.M. Ingram, M.F. Perutz: Proc. Roy. Soc. London *A225*, 287 (1954)
2.2 W.L. Bragg, M.F. Perutz: Proc. Roy. Soc. London *A225*, 315 (1954)
2.3 G. Bodo, H.M. Dintris, J.C. Kendrew, H.W. Wyckhoff: Proc. Roy. Soc. London *A253*, 70 (1959)
2.4 C.C. Phillips: In *Advances in Structure Research by Diffraction Methods*, Vol.2, ed. by R. Brill, R. Mason (Vieweg, Braunschweig 1966) p.75
2.5 W. Hoppe: Isr. J. Chem. *10*, 321 (1972)
2.6 F. Parak, R.L. Mössbauer, C. Hermes: Eriwan Proceedings, to be published
2.7 A. Kastler: Compt. Rend. Acad. Sci. *250*, 509 (1960)
2.8 P.B. Moon: Nature *185*, 427 (1961)
2.9 R.S. Raghavan: Proc. Ind. Acad. Sci. *53*, 265 (1961)
2.10 P.J. Black, D.E. Evans, D.A. O'Connor: Proc. Roy. Soc. London *270A*, 168 (1962)
2.11 R.N. Kuz'min, A.V. Kolpakov, G.S. Zhdanov: Sov. Phys. Crystallogr. *11*, 457 (1967)
2.12 G.S. Zhdanov, R.N. Kuz'min: Acta Crystallogr. *B24*, 10 (1968)
2.13 P.J. Black, G. Longworth, D.A. O'Connor: Proc. Phys. Soc. London *83*, 925 (1964)
2.14 P.J. Black, I.P. Duerdoth: Acta Crystallogr. *21*, Pt 17, Suppl.A214 (1966)
2.15 V.K. Voitovetskii, I.L. Korsunskii, A.I. Novikov, Yu.F. Pazhim: JETP Lett. *54*, 1361 (1968)
2.16 A.M. Artem'ev, I.P. Perstnev, V.V. Sklyarevskii, G.V. Smirnov, E.P. Stepanov: JETP Lett. *64*, 261 (1973)
2.17 F. Parak, R.L. Mössbauer, U. Biebl, H. Formanek, W. Hoppe: Z. Phys. *244*, 456 (1971)
2.18 F. Parak, R.L. Mössbauer, W. Hoppe: Ber. Bunsenges. Phys. Chem. *74*, 1207 (1970)
2.19 R.L. Mössbauer: Naturwissenschaften *60*, 493 (1973)
2.20 R.L. Mössbauer: In *Anomalous Scattering*, ed. by S. Ramaseshan, S.C. Abrahams (Munksgaard Internat. Copenhagen 1975) p.463
2.21 F. Parak: In *Proc. Int. Conf. Mössbauer Spectroscopy*, Vol.2, ed. by A.Z. Hrynkiewicz, J.A. Sawicki (Cracow 1975) p.285
2.22 B.P. Schoenborn: Nature *224*, 143 (1969)
2.23 B.P. Schoenborn, A.C. Nunes, R. Nathans: Ber. Bunsenges. Phys. Chem. *74*, 1202 (1970)
2.24 W. Hoppe, U. Jakubowski: In *Anomalous Scattering*, ed. by S. Ramaseshan, S.C. Abrahams (Munksgaard Internat. Copenhagen 1975) p.437

2.25 W. Hoppe: private communication
2.26 R.L. Mössbauer, F. Parak: to be published
2.27 G.T. Trammell: Phys. Rev. *126*, 1045 (1962)
2.28 J.P. Hannon, G.T. Trammell: Phys. Rev. *186*, 306 (1969)
2.29 Yu. Kagan, A.M. Afanas'ev: Z. Naturforsch. *28a*, 1351 (1973)
2.30 R. Zimmermann: Nucl. Instrum. Methods *128*, 537 (1975)
2.31 Y. Maeda, T. Harami, A. Trautwein, U. Gonser: Z. Naturforsch. *31b*, 487 (1976)
2.32 F. Parak, U.F. Thomanek, D. Bade, B. Wintergerst: Z. Naturforsch. *32c*, 507 (1977)
2.33 R.L. Mössbauer: J. Phys. Paris Colloq. C6 *37*, C6-11 (1976)
2.34 F. Parak, R.L. Mössbauer, W. Hoppe, U.F. Thomanek, D. Bade: J. Phys. Paris, Colloq. C6 *37*, C6-703 (1976)
2.35 J.G. Mullen, H.N. Ok: Phys. Rev. *168*, 550 (1968)
2.36 H.N. Ok, W.R. Helms, J.G. Mullen: Phys. Rev. *187*, 704 (1969)
2.37 J.G. Mullen, H.N. Ok: Mössbauer Eff. Methodol. *4*, 103 (1968)
2.38 W.R. Helms, J.G. Mullen: Phys. Rev. *B4*, 750 (1971)
2.39 J. Loock, R.N. Kuz'min, F. Parak: to be published
2.40 F. Parak, E.N. Frolov, R.L. Mössbauer, V.I. Goldanskii: J. Mol. Biol. *145*, 825 (1981)
2.41 F. Parak, H. Formanek: Acta Crystallogr. *A27*, 573 (1971)
2.42 U.F. Thomanek, F. Parak, R.L. Mössbauer, H. Formanek, P. Schwager, W. Hoppe: Acta Crystallogr. *A29*, 263 (1973)
2.43 R. Huber, D. Kukla, W. Bode, P. Schwager, K. Bartels, J. Deisenhofer, W. Steigemann: J. Mol. Biol. *89*, 73 (1974)
2.44 J. Deisenhofer, W. Steigemann: Acta Crystallogr. *B31*, 238 (1975)
2.45 H. Frauenfelder, G.A. Petsko, D. Tsernoglau: Nature *280*, 558 (1979)
2.46 S. Glauser, M.G. Rossmann: Acta Crystallogr. *21*, 175 (1966)
2.47 W. Longley: J. Mol. Biol. *30*, 323 (1967)
2.48 B.K. Vainshtein, V.V. Barynin, G.V. Gurskaya, V.Ya. Nikitin: Sov. Phys. Crystallogr. *12*, 750 (1968)
2.49 B.K. Vainshtein, V.V. Barynin, G.V. Gurskaya: Sov. Phys. Dokl. *13*, 838 (1969)
2.50 Rich.A. McPherson: Arch. Biochem. Biophys. *157*, 23 (1973)
2.51 W. Eventoff, G.V. Gurskaya: J. Mol. Biol. *93*, 55 (1975)
2.52 W. Eventoff. N. Tanaka, M.G. Rossmann: J. Mol. Biol. *103*, 799 (1976)
2.53 A.L. Marie, F. Parak, W. Hoppe: J. Mol. Biol. *129*, 675 (1979)
2.54 A.L. Marie, H. Prieß, F. Parak: Hoppe Seylen Z. Physiol. Chem. *359*, 857 (1978)
2.55 U. Biebl, F. Parak: Nucl. Instrum. Methods *112*, 455 (1973)
2.56 G. Charpak, Z. Hajduk, A. Jeavons, R. Stubbs: Nucl. Instrum. Methods *122*, 307 (1974)
2.57 G. Charpak, C. Demierre, R. Kahn, J.C. Santiard, F. Sauli: Nucl. Instrum. Methods *141*, 449 (1977)

# 3. The Gravitational Red-Shift

R. V. Pound

**With 3 Figures**

The discovery of emission and absorption of $\gamma$ rays free of thermal Doppler broadening and the effects of recoil by R.L. MÖSSBAUER [3.1] brought to experimental physics a tool that has been used to detect fractional changes of electromagnetic energy smaller than can be detected by any other technique. One of the earliest and most striking applications was the confirmation of the effect of gravity on photon energy as predicted by A. Einstein. The general background and some details of the most thorough of such studies will be described herein.

## 3.1 Mössbauer's Discovery

The $\gamma$-ray resonance first reported by MÖSSBAUER [3.1] displayed a width in energy approximately $10^{-10}$ times its mean energy. Correspondingly, the Doppler effect resulting from movement of the source toward or away from the absorber at speeds of 1 cm/s, or $0.3 \times 10^{-10}$ times c, the velocity of light, sufficed to reduce the absorption to one-half its maximum value. Physicists in several parts of the world independently recognized, soon after they learned of Mössbauer's discovery, that the phenomenon might be developed to detect and measure the effect of differing gravitational potentials on frequency or time as predicted by EINSTEIN in 1907 [3.2]. There came to the notice of the present author, soon after his own joint proposal with REBKA had been published [3.3], word of similar plans by DEVONS, BOYLE and BUNBURY of Manchester University [3.4], by CRANSHAW (as reported by Schiffer and Marshall at Harwell) [3.5], by DICKE at Princeton [3.6], by BARIT, PODGORETSKII and SHAPIRO at Dubna [3.7], and by SHIMODA in Tokyo [3.8].

### 3.1.1 The Principle of Equivalence and the Red-Shift

EINSTEIN had formulated his Principle of Equivalence (P of E) in 1907 and 1911 [3.2] and had derived from it the prediction of an astronomical gravitational shift of spectral lines toward the red. The purpose of the P of E was to enable effects of gravitation to be analyzed within the framework of Special Relativity, as contained in the Lorentz transformation. The concept was an important step toward the formulation of the theory of gravitation as contained in the General Theory of Relativity.

According to the P of E, in a local region it is not possible to distinguish between effects of a uniform gravitational field and effects that would be found in a reference frame accelerating relative to an inertial frame at the rate of, and in the direction opposite to, the acceleration of objects considered to be in free fall. This construct leads very easily to the prediction of a red-shift of spectral lines coming from massive sources. Suppose an observer studies a source situated above him a small distance h in an earthbound laboratory where the local acceleration due to gravity is g. By supposing the laboratory instead to be accelerating upward at the rate g one concludes that in the time h/c taken for light to travel from source to observer, the observer has increased his speed toward the source by gh/c. Consequently he sees the light fractionally shifted toward the violet as a Doppler effect in the first order of $gh/c^2$. This shift amounts to $1.09 \times 10^{-16}\ m^{-1}$ at the surface of the earth and is so small that the effect was thought detectable only in the far stronger gravitational fields of stars. In that case the quantity gh is generalized to the gravitational potential difference $\Delta\Phi \approx GM_s/R$, where G is the gravitational constant, $M_s$ is the mass, and $R_s$ the radius of the star. The gravitational effect of the earth has been ignored as negligible in this case. This shift would be toward the red and in the fractional amount $\Delta\Phi/c^2 = GM/Rc^2$.

### 3.1.2 Astronomical Studies

For the sun, the star best understood by the astronomers, the fractional shift should result in a wavelength increase $\Delta\lambda/\lambda_0 = 2.12 \times 10^{-6}$, when solar lines are compared to the same lines generated on earth. In the years between 1911 and 1959 many studies of solar spectra were carried out with the aim of testing the validity of the predicted red-shift [3.9-12]. Nevertheless, the evidence remained inconclusive largely because the effect sought was so small compared to the breadths of most Fraunhofer lines. Results for most solar lines studied depended upon the point of origin of the line on the solar disk. The majority of the lines catalogued by ADAM [3.11] and by FINLAY-FREUNDLICH [3.12] showed shifts only about 1/3 as large as predicted when originating near the center of the solar disk. The shifts increased and possibly exceeded the predicted value as the origin was moved to the limb. The line profiles were found to lack symmetry, the shapes changed between the center and the limb, and the widths were generally several times larger than the predicted shifts. All of these properties made evaluation of the shift difficult. An example of the line shape, linewidth, and the problem of determining the shift is illustrated by BLAMONT and RODDIER [3.13] who studied a solar line of strontium with a resonant atomic beam technique in 1961. More recently still, BRAULT [3.14] and SNIDER [3.15] have studied alkali lines reported by St. JOHN to be relatively symmetric and to have little variation from the center to the limb [3.10]. These lines are presumed to be formed high in the solar atmosphere. Agreement with the prediction within 5% for a sodium line was reported by Brault who used an electronic photodetection technique and to about the same level for potassium by Snider, using an atomic beam as a resonant detector.

Effort has also been directed to studies of light from stars with much larger predicted shifts, such as white dwarfs with larger values of $M_S/R_S$. It is hard to see how a conclusive result could be obtained from such sources with so much less detailed knowledge of their structures than for the sun.

A short time before the Mössbauer effect became known, the development of clocks having much improved stability resulted in proposals to test the prediction using atomic clocks in space [3.16]. The contribution to the change of rate because of the change of gravitational potential for an orbit of radius many times that of the earth should be an increase by $7 \times 10^{-10}$. This is well within the stability of clocks based on cesium or hydrogen H.F.S. or even quartz crystals. Time dilation of scale similar to that and Doppler shifts many orders of magnitude larger are introduced by the motion of a space vehicle. Considerable ingenuity is required to compensate for these [3.17].

### 3.1.3 Application of the Mössbauer Effect

The Mössbauer effect, in offering a resolution sufficient to measure the effect of gravity in a ground based environment, carries with it the especially distinctive quality of allowing the correlation between the shift observed and the relevant gravitational potential difference to be explored explicitly. For example, the sign of the effect can be reversed by inverting the sense of travel over a fixed vertical path. In contrast, the astronomical tests are observations coupled with reasonable, but to some extent uncertain, assumptions as to the properties of the observed system.

Interest in exploitation of the Mössbauer effect to resolve the uncertainties about the gravitational red-shift led to a search for an example of the phenomenon that maximized the sensitivity to small fractional frequency shifts. If one envisages an experiment based on the conventional transmission of $\gamma$-rays from a source through an absorber with the two elements separated by a vertical path of height h, it is easy to see that for a given source strength, the parameters needed are narrow fractional width and large fractional absorption depth F. The fractional width can be described by $V_H$, the source speed that reduces the absorption to 0.5 times its maximum value. If no causes of line broadening were present other than the decay time, one would seek to develop resonances of maximum excited-state lifetime and highest $\gamma$-ray energy consistent with a large resonant depth F. Interest in such an extreme example as $^{109}$Ag, with lifetime 58 s and energy 88 keV has been expressed for the two decades [3.6,18] since the early work, but so far useful results have not been realized.

The interaction of nuclear magnetic moments among one another, imperfections in the crystal structure that give rise to a stochastic spread of crystal fields and through them of electric quadrupole splittings, and variations in the isomer shift from chemical and physical inhomogeneity all contribute to prejudice the realization of arbitrarily sharp lines. In practice, the experience from radiospectroscopy of nuclear magnetism in the solid state provides evidence of the range of line widths one might encounter as a lower limit. Widths less than a kilohertz are not easily achieved in the solid state, especially for nuclei of spin greater than 1/2, where inhomogeneous quadrupole splitting may contribute.

Perusal, in 1959, of the tables of isotope properties then known led to the identification of three particularly interesting candidates. These were the 14.4-keV transition from a first excited state of lifetime 0.1 μs in $^{57}$Fe, the 93-keV transitions from a 9.4-μs state of $^{67}$Zn, and a 13.3-keV transition from the 2.98-μs state of $^{73}$Ge. The fractional width of the resonance for $^{57}$Fe should ideally be $7\times10^{-13}$; for $^{67}$Zn, $10^{-15}$; and for $^{73}$Ge, about $3\times10^{-14}$. Several longer-lived isomeric transitions offered tantalizing possibilities of resonance lines fractionally much narrower, but little progress has been made toward observing such resonances. The example of $^{109}$Ag was mentioned above.

One other isomeric transition having parameters that made it of great interest for this application was discovered shortly after the resonance of $^{57}$Fe had been explored. This was a 6.25-keV transition from a first excited state of half-life 6.8 μs in $^{181}$Ta [3.19]. In this example, a fractional width of about $2.5\times10^{-14}$ is available in principle. As for $^{57}$Fe and $^{73}$Ge, the small recoil energy associated with the low γ-ray energy and the reasonably stiff lattices in which it occurs should result in large recoil-free fractions even at room temperature. On the other hand, an internal conversion coefficient of 45 renders the 6.25-keV γ ray of $^{181}$Ta difficult to separate from copious X rays at 8 and 50 keV.

It was not until Si(Li) γ-ray spectrometers became available that either the ray itself, as distinct from the conversion electron, or the resonance in $^{73}$Ge was detected [3.20]. With a conversion coefficient of 1100 [3.21], counting rates in the desired γ ray of $^{73}$Ge are restricted and absorption cross sections are small, even though the natural line width has been achieved [3.22]. In contrast, the internal conversion coefficient in $^{57}$Fe of 9.0 seems small although even there attention must be paid to the presence of much more copious X rays at 6 keV and the little converted intense γ ray at 122 keV.

So far, $^{57}$Fe remains the isotope best suited to most applications aimed at detecting small energy changes. In the discussion of the applications to follow, the work of POUND and REBKA [3.23] and of POUND and SNIDER [3.24] will be described in some detail.

## 3.2 Modulation Technique to Detect Small Shifts

To apply the resonance of $^{57}$Fe to the measurement of a very small change in the relative frequencies of a source and an absorber caused by a change in the sense of gravity, one seeks a combination of the largest resonant depth of absorption with a minimum of nonresonant absorption. It is now known that copper or palladium make excellent matrices to carry, as a dilute impurity, the carrier-free 270-day parent $^{57}$Co. Either of these source matrices yields an unsplit emission line of near the minimum breadth. Unfortunately, no unsplit absorber matched to such a source is available. POUND and REBKA [3.23], POUND and SNIDER [3.24] and CRANSHAW, SCHIFFER and WHITEHEAD [3.25] used soft iron for both source and absorber. The presence of the internal magnetization and its attendant hyperfine structure, even if perfectly matched between source and absorber, weakens the absorption cross section at the central maximum relative to an unsplit combination, although line widths nearly as small as permitted by the lifetime can be realized. In order to obtain good absorption depth without large nonresonant absorption, isotopic enrichment of the absorber is of considerable benefit. One radiation length for nonresonant absorption is 15 mg/cm$^2$. In the experiment of POUND and SNIDER the absorber was 1.5 to 2.0 mg/cm$^2$ enriched to 43% in $^{57}$Fe. Thus, nonresonant absorption was less than 15%. In order to make observable the offset of frequency of the source relative to the absorber under a given arrangement, two different motions were imposed on the source. First, the source was vibrated sinusoidally at about 74 H toward and away from the absorber. The vibration was induced by a hollow cylindrical ceramic ferroelectric transducer by applying a modulating sinusoidal voltage across the thin (1 mm) transducer wall. The velocity extrema of this motion corresponded approximately to the points in the resonance curve where the greatest rate of change of absorption with velocity for the source-absorber combination were recorded. Counts, received in the quarter cycles of the oscillation period centered around each of the two extrema, were directed to different registers for accumulation. Thus one register recorded transmission on the side of the absorption line when the modulation velocity was directed toward the absorber and the other when the velocity was directed away from the absorber. A second very slow motion was also applied to the source. This motion was produced by the piston of a slave hydraulic cylinder on which the source and ferroelectric transducer were mounted. The piston was caused to move smoothly at a velocity very small compared to the velocity width of the absorption line by moving a smaller-diameter master cylinder by a screw drive from a synchronous clock motor. This motion, at a speed less than $10^{-3}$ cm/s was sufficient to introduce a Doppler shift large compared to the effect being sought but still small compared to the line width. It was maintained in one direction for a fixed integral number of vibrational cycles and then reversed for an equal number. The counts for each part of the vibrational period were routed to different counters for each of the two signs of the slow, or calibrating, motion introduced hydraulically. There were, thus, four registers accumulating data on the main absorber and detector.

From the four counts obtained during a given running time, the velocity corresponding to the apparent maximum absorption could be deduced as a fraction of the velocity imposed by the hydraulic piston. The calibration of the system involves only that quantity, which was measured carefully on several occasions using a precision dial gauge to observe the net displacement in runs lasting approximately five minutes. In effect, the combination of the two signs of hydraulic motion and the two of vibration results in transmission measurement at two points on each side of the absorption line near the point of steepest slope. One can visualize the determination of the line center as the intersection point between the straight lines drawn through the two points on each side. A change in that apparent line center by the inversion of the system is then found by comparing data reduced in the same way with the inverted system.

If the line shape is assured to be Lorentzian, centered nominally at zero velocity, the greatest sensitivity to small displacements is obtained if the vibrating velocity is $\sqrt{3}\, V_H/3$, where $\pm V_H$ are the velocities that reduce the absorption to half its maximum value. Statistical uncertainty in the velocity deduced for the line center arises from the finite counting rate available. Expressed as a fractional uncertainty of frequency, the standard deviation of the fractional displacement of the absorption line from that of the source becomes

$$\delta a_0 = \left[\left(1-\frac{3}{4}F\right) / \, 3N_0\right]^{\frac{1}{2}} \times 8V_H/3Fc \quad , \tag{3.1}$$

where F is the fraction by which the intensity is reduced at the maximum of the resonant absorption, c is the velocity of light and $N_0$ is the total number of counts that would have been detected in the absence of the resonant absorption. In terms of the time of a run $\tau$ with a source that suffers R decays per unit time,

$$N_0 = R\tau b\varepsilon D\Omega/4\pi \quad ; \tag{3.2}$$

where b is the fraction of decays that yield the resonant $\gamma$ ray; $\varepsilon$ is a factor less than unity allowing for detector inefficiency and nonresonant absorption; D is a duty cycle for the modulation scheme, nominally about 0.5 in the scheme described above; and $\Omega/4\pi$ is the solid angle subtended at the source by the absorber and its detector.

It is clear from (3.1) that $F/V_H$ can be regarded as an effective figure of merit for this sort of application of the Mössbauer effect, in so far as statistical limitations apply. An important factor favoring $^{57}Fe$ is the large value of F that can be easily obtained. For example, a fractional depth of about 0.3 is available without serious absorption broadening even at temperatures elevated somewhat above normal room environments. With $V_H$ near 100 µm/s the value of $F/V_H$ of $3\times10^{-3}$ is obtained.

In comparison, the very narrow line of $^{67}Zn$ should yield $F/V_H = 0.01/0.15 = 6 \times 10^{-2}$, about 20 times greater. In exchange for that gain, however, are the facts that low temperatures must be employed to realize such a value of F and that the lifetime of only 76 hours of the $^{67}Ga$ parent sets a small limit to the running time available without continuous renewing of sources. Another difficulty arises from the need to employ isotopically enriched material as the absorber to realize a fraction F anywhere near 0.01. Such material exists in only small amounts. Because of the shorter wavelength and small recoil free fraction of $^{67}Zn$, an amount of the order of magnitude of 1 g/cm$^2$ of absorber area is needed as compared with about 1 mg/cm$^2$ for an $^{57}Fe$ system. For an absorber of the area employed in the $^{57}Fe$ experiment, if it were available, the enriched $^{67}Zn$ would cost about \$$4.5 \times 10^6$ at present Oak Ridge prices. Such a large system would obviously be prohibitively expensive. There is very important practical difference between a system based on the narrow but shallow $^{67}Zn$ resonance as compared to that based on $^{57}Fe$. For the technique employed to integrate data over long periods, vibrations that introduce source-to-absorber Doppler shifts smaller than a reasonable fraction of the linewidth can be present without seriously affecting the behavior of the system. So long as the vibrations remain within the part of the line of reasonably constant slope, little rectification occurs, and only drifts averaged over the run contribute to the results obtained over a given period. With a narrow line, such as that of $^{67}Zn$, sensitivity to spurious vibrations is correspondingly 1000 times greater than it is for $^{57}Fe$. So far, all applications of the narrow $^{67}Zn$ resonance have involved small, almost rigid, source and absorber structures. Separation of source and absorber by large distance would entail a serious structural problem to achieve adequate rigidity and stability.

One final point should be remarked in applying (3.1,2). The fractional uncertainty for a given running time and for a given linear dimension of absorber is proportional to the source-to-absorber distance h through the factor $N_0^{-\frac{1}{2}}\alpha\Omega^{-\frac{1}{2}}\alpha h/d$. However, the predicted effect of gravity is also proportional to h, and as a result the statistical uncertainty becomes independent of h. Nonetheless, the need to control several sources of serious systematic error favors the use of a reasonably large value of h.

### 3.2.1 The Enclosed Tower of the Jefferson Physical Laboratory

In the experiments carried out at Harvard between 1959 and 1964, the vertical baseline was provided by an isolated tower structure internal to the Jefferson Physical Laboratory (see Fig.3.1). The tower was a special feature of the building since its construction in 1884. This structure forms some interior corridor walls and extends from a subterranean room, originally used to provide an environment of constant temperature, to a penthouse structure above the central highest part of the roof. In recent years the separation between the tower walls and the main bearing walls of the building was bridged in a limited number of places. As a consequence, the isolation from vibration arising from activities in the building was less complete than

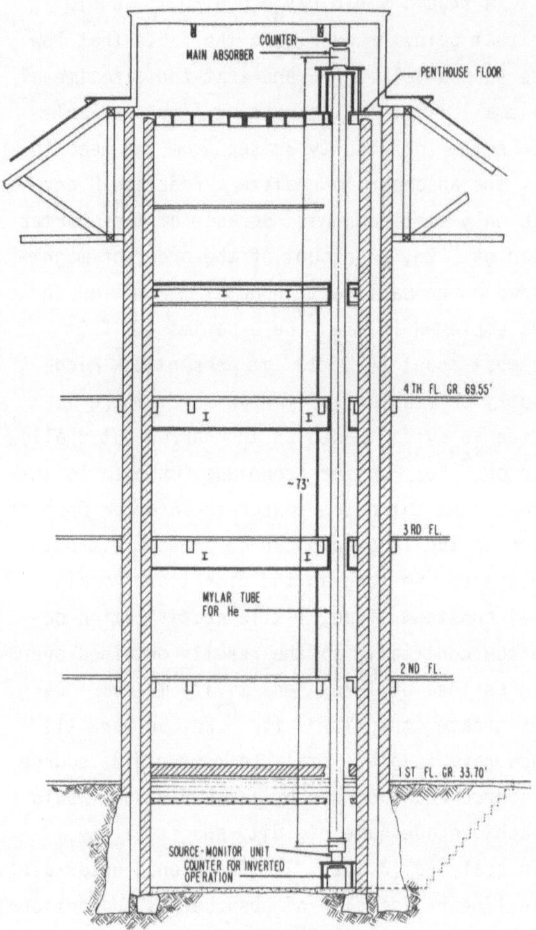


Fig. 3.1. The configuration of the experiment in the "tower" at the Jefferson Physical Laboratory

it might have been. It was necessary to eliminate air from the path of the 14.4-keV $\gamma$ rays because its mean absorption distance in air is only about 5 m. A path through the seven intermediate floor levels, with an absorption much less than air, was provided by installation of a length of thin-walled mylar tubing inflated with helium gas. The tubing was supported on brass flanges at each end, closed by mylar end windows, and purged of heavier gases that might diffuse or leak in by the slowly flowing helium, in at the top and exhausted out the bottom. As a result, the $\gamma$-ray absorption in the path was held to about 12%, corresponding to the mean absorption distance in helium at STP of about 200 m.

The pulses transmitted through the main absorber were detected in either of two large proportional counters. These used a single central wire and a metal foil and mylar entrance window and were filled with an argon-methane mixture. One counter was located on the floor at the bottom of the tower and the other on the ceiling of the penthouse at the top.

### 3.2.2 The Effect of Temperature

In the course of operating the original system in 1960, Pound and Rebka were able to correlate an instability in results with changes of temperature and were led to the discovery of the effect of temperature on the energy of the Mössbauer $\gamma$ rays. They explained the effect of temperature changes as arising mainly from the time dilation, or "second-order relativistic Doppler effect" due to the thermal vibrations of the atoms in the lattice [3.26]. The binding of the nuclei to the lattice sites causes the velocity of thermal vibration v, and so the ordinary Doppler effect, to vanish when averaged over such long times as the nuclear $\gamma$-ray lifetime, but $<v^2>_{Av}$ is directly proportional to the thermal excitation and zero-point energies. This phenomenon was explained independently and almost simultaneously by JOSEPHSON [3.27]. There are other terms contributing to a temperature dependence, for example, a variation of the isomer shift from thermal expansion [3.28], but the largest component is the time dilation. In the early experiment, measurements of the temperature difference between the source and absorbers was provided by thermocouples with one junction thermally attached to each and the e.m.f. was recorded continuously. Data on frequency shifts were corrected to equal average temperatures for each run, using the calculated temperature coefficient of the fractional frequency shift of $-2.21 \times 10^{-15}/^{\circ}C$ which was consistent with the best measurement of that time. In the more refined experiment of Pound and Snider, all of the elements were mounted in temperature controlled ovens and thermistors provided continuous monitoring of the temperature. In this case the temperature coefficient in the region of the $43.5^{\circ}C$-oven temperature was measured by observing the frequency shift resulting from $\pm 5^{\circ}C$ temperature changes. The resulting coefficient of $-(2.12 \pm 0.02) \times 10^{-15}/^{\circ}C$ was then used to correct for the measured small variations in the average controlled temperatures from run to run.

### 3.2.3 The Monitor Channel and Source Unit

Emphasis has been put on the role played by the rate of change of transmission through the absorber with frequency, or Doppler velocity, rather than the linewidth itself, as the factor determining the feasibility of this application. With $^{57}Fe$, the shift expected in the 22 m height, $2.5 \times 10^{-15}$ fractionally, was only $2.5 \times 10^{-3}$ parts of the linewidth. Statistical limitations could be shown to allow a measurement of such a shift to well under 1% if data were averaged for a run of several weeks. The effect due to gravity could only be separated from other source-absorber shifts by correlating it with inversion of the system, but precautions were needed to avoid some obvious systematic errors associated with the inversion. For example, the transducer used to generate audio-frequency vibrations might be expected to perform differently in its two orientations, rightside up and upside down. To deal with that possibility, a complete system of absorbers, proportional counters, and electronic gates and storage registers was added to provide a continuous measurement of the absorption line asymmetry observed from about 30 cm away from the source. In the final version of the experiment there were two such "monitor" absorbers mounted within a common temperature-controlled housing with the source and its transducers sampling the $\gamma$-ray beam just outside the angle intersecting the remote main absorber. Two absorbers and detectors were used on diametrically opposite sides of the beam, and apertures were adjusted so that they contributed approximately equally to the superposed net counting rate. This design was made as a precaution against the possible presence of a motion of the transducer in the transverse direction to which a single-sided monitor channel could have been sensitive. With this monitor system, apparent line displacements were computed for each run in just the same way as for the main absorber channel. Thus, the final result attributable to gravity was the difference of the difference between the main and monitor line positions that accompanied the inversion. A test of the performance of this monitor arrangement was carried out by taking data with strongly distorted waveforms applied to the ceramic transducer. Although apparent line displacements as large as ten times the signal expected from gravity were introduced, as measured independently on the main or the monitor channel, the difference remained constant to well within statistical limits, which were perhaps 10% of the gravitational signal. Data taken in actual runs did not reveal shifts in the monitor data with inversion on a scale as large as 1/40 of those applied in this test. So it appeared that the functioning of the monitor system was adequate to eliminate this source of error.

Another precaution taken against spurious effects associated with the system inversion was the use of polarizing magnetic fields fixed to the source and all absorber units. These were provided by ceramic permanent magnets and the fields were in the plane of the source and absorber foils. The field achieved in the main absorber was about 50 Oe and in the source and monitor absorbers it was greater than 100 Oe. In each operation of the system the units were put into place with their magnetic fields parallel, with a consequent increase in absorption depth. It was supposed safe to ignore the necessary changes of orientation of the system with respect to the earth's magnetic field, because the inherent fields in the units were at least one hundred times larger.

The source unit employed in the final version of the experiment included a slave hydraulic cylinder, on the piston rod of which was mounted the 10 cm long, 4 cm diameter ferroelectric ceramic transducer. The source itself consisted initially of 1.25 Ci of $^{57}$Co diffused at 1050$^{\circ}$C for 24 h into a foil of thickness 5 mg/cm$^{2}$ enriched to 99.7% $^{56}$Fe in order to avoid resonant absorption in the source. The source was made as large as 10 cm in diameter to limit the alloying effects from the Co impurity added. Such an impurity could affect the hyperfine structure and therefore the apparent linewidth when used with pure iron absorbers. The source foil was cemented to a stiffened copper plate about 3 mm thick, which was epoxied to the end of the ceramic transducer. A precision thermistor was glued to the source as a temperature sensor and its resistance was monitored continuously on a chart recorder with a resolution of approximately 0.001$^{\circ}$C.

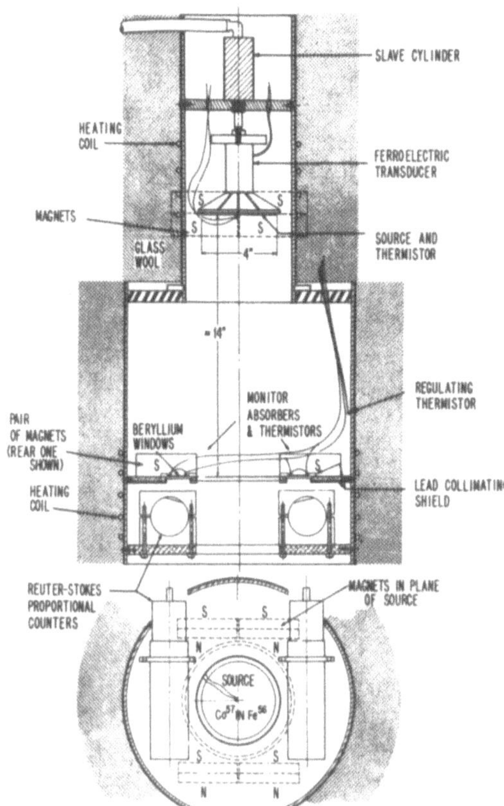


Fig. 3.2. Two sections through the combined source and monitor unit. In the upper view a vertical cut shows the hydraulic piston, transducer, source, the monitor absorbers, and proportional counters all inside the thermally regulated oven. The lower view is a horizontal cut throught the monitor counters and shows the magnets polarizing the source

Attached rigidly to the heated aluminum cylinder housing the source and transducer structure was a larger cylinder that provided the thermal environment and mountings for the monitor absorbers and their related proportional counters (Reuter-Stokes RSG-30A filled with a krypton-nitrogen gas mixture at STP). A thermistor was cemented to each monitor absorber and a continuous recording made of the mean temperature of the pair. Heater coils were wound around each part of the combined monitor and source unit. The temperature, sensed by yet another thermistor in the walls of the unit, was held constant through a temperature regulating bridge that provided proportional control of the power in the heater coils. The entire unit was jacketed in a layer of lead to reduce the radiation level nearby, except in the beam to the main absorber. Thermal lagging was provided by wrapping in glass-wool blanket. A drawing of two cross sections of the combined source-monitor unit appears in Fig.3.2.

### 3.2.4 The Main Absorber

The main absorber was composed of a mosaic of foils each approximately 5 cm square on a side, rolled from iron enriched to 43.5% $^{57}$Fe. These foils were cemented to a Mylar sheet 0.15 mm thick. They were of approximately 1.8 mg/cm$^2$ thickness and covered the area of a circle 38 cm in diameter. Any gaps between foils were covered with lead paint to reduce such contribution to the background. The Mylar sheet was cemented to a honeycomb structure (Hexcel) of aluminum material 0.08 mm thick with cells about 2 cm in diameter and 2.5 cm long. This support provided some transverse heat conductivity and rigidity with a minimum obstruction of the $\gamma$-ray flux. This structure was placed in the midplane of an aluminum tube 40 cm inside diameter and 45 cm long, forming an oven. The central 15 cm was flanged with O-rings to grip transverse Mylar windows above and below the absorber foil, allowing evacuation of that part of the unit. This was done mainly to reduce the impact of airborn noise on the absorber diaphragm. A series of three more thin plastic diaphragms, the outer of which were aluminum coated, were installed across the aperture above and below the vacuum windows in order to reduce the thermal effects of air currents. Heater coils were wound around the upper and lower sections of the 40 cm cylinder and a heating element wound on a thin plastic cylinder that caused little obstruction to the $\gamma$ rays was placed between the outermost plastic diaphragms at each end. Temperature was controlled by reference to thermistors mounted on the cylinder wall. Six more thermistors were cemented to the main absorber foil at different azimuths and radii to allow the assessment of the thermal homogeneity of the absorber. Data from

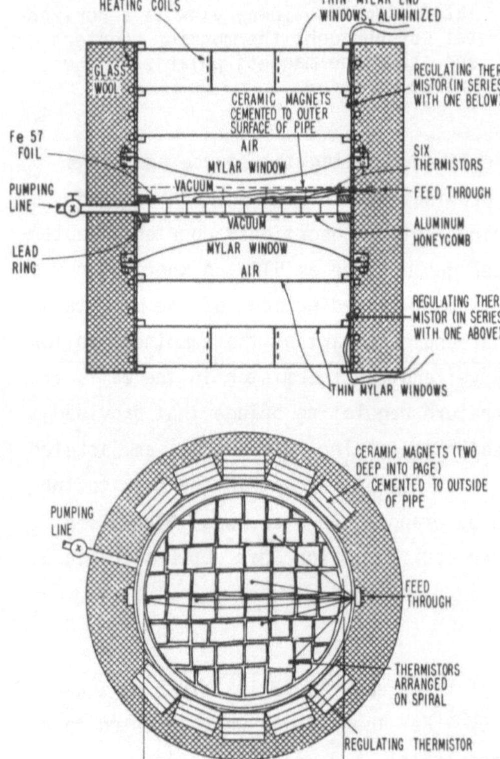


Fig. 3.3. Two sections through the main absorber oven. The lower view is an horizontal cut that shows the mosaic of absorber foils and the stacks of permanent ceramic magnets that provided a polarizing field

one of these thermistors was recorded continuously during all runs when data were being taken. The magnetic field to polarize the absorber foils was provided by stacks of ceramic permanent magnets cemented around the midplane of the oven, creating a transverse field of 50 Oe or more in that plane, and not varying greatly in direction. The oven was also wrapped in glass wool for thermal insulation. Cross sectional drawings of the main absorber unit appear in Fig.3.3.

## 3.3 Data Taking and Results

Each run of data taking yielded eleven numbers, i.e., eight counts of $\gamma$ rays and three average temperatures (thermistor resistances). Counts for the four main absorber registers and the four monitor registers were usually recorded after every 12 full cycles of the calibrating hydraulic piston motion. That motion was reversed every 22,000 periods of the 74 Hz vibration of the source. Three continuous chart recordings were made of the unbalance of each of the three Wheatstone bridges associated with the temperature sensing thermistors on the main and monitor absorbers and the source. For each running period the offset in fractional frequency of the main absorber from the source, of the monitor from the source, and of the main absorber from the monitor was calculated, including corrections to constant average temperatures. The source-monitor unit and the main absorber were interchanged every few days to allow the comparison of the results from a rising $\gamma$-ray beam to those with a falling one, with about equal weights. In order to eliminate uncertainty about the effect of inverting the main absorber oven, it was operated for about equal times at each end of the system in each of the two inversions and data were averaged over these, even though any systematic difference found in the data for the two orientations separately was not statistically significant.

All data were reduced to effective Doppler speeds divided by c, or fractional frequency shifts. The change of the difference between the main channel and monitor channel displacements that accompanied the inversion was found to be $(4.902\pm0.041) \times 10^{-15}$ from data taken over about four months of operation. This may be compared to the value $4.905\times10^{-15}$ calculated from $2\ hg/c^2$ with $2h=44.96$ m and g given the tabulated local value, 9.804 $m/s^2$. In addition to these runs, data were taken for about one month using only the lower one-half (approximately) of the available tower height. Here again, both rising and falling $\gamma$ rays were observed, and the absorber oven was operated both ways-up at each end. The result from these runs was $(2.195\pm0.040)\times10^{-15}$, which is to be compared to the predicted value of $2.226\times10^{-15}$ for the height used. When all of these measurements are combined, they yielded a result $0.9970\pm0.0076$ times the result predicted from the principle of equivalence.

### 3.3.1 Systematic Errors

In all of the results quoted so far the uncertainties given represent standard deviations arising only from finite counting statistics. Allowance must also be made for the possible presence of systematic errors. Among the items that contribute uncertainties of this kind are the limited precisions in the measurements of the path length, of the calibration velocity, and of the temperature coefficient of the $^{57}$Fe resonance. There was undoubtedly some error associated with the nonlinearity of the vibration-based differentiating process over the range of the calibrating velocity, but the largest and dominant error was thought to arise from the inexact representation of the weighted mean temperature by the temperature sensed and recorded continuously at one thermistor on the absorber. A test was made, after the main runs had been completed, to see how similar to one another were the changes at the six thermistors when the reference temperature was changed by ±5°C from the normal value. It was assumed that such a test would reveal discrepancies similar to those accompanying changes of ambient temperature with fixed reference temperature at 43°C. The result lead to the conclusion that as much as ±0.005 fractional error could have been introduced by this source alone. Altogether the combination of this and the other systematic errors with the statistical ones, taking them all to be statistically independent, results in a final result well within an overall uncertainty of ±0.01.

## 3.4 Other Recent Red-Shift Experiments

Since this experiment was carried out, several experiments based on other techniques have contributed to the verification of the prediction from the principle of equivalence of the effect of gravitational potential difference on photon energy and on time scales. The solar observations of BRAULT and of SNIDER were mentioned earlier. HAFELE and KEATING [3.29] compared the effect on the time keeping of cesium atomic clocks by transporting them in easterly and westerly directions using commercial jet aircraft. It is difficult to establish the level of overall error for the part of the effect measured attributable to gravity but overall consistency with theory was found. BRIATORE and LESCHIUTTA [3.30] in 1977 reported measurements of time differences derived from a cesium beam atomic clock operated at an altitude of 3500 m in the Italian Alps and one operated at Turin, Italy, at 250 m altitude. They found agreement with expectation with an uncertainty of about ±0.15. ALLEY and others [3.31] compared atomic clocks maintained in flight for fifteen hours at altitudes ranging from 8 to 11 km, with identical ones on the ground. Their result, after correcting for time dilation effects from the motion, yielded 0.987 ± 0.016 times the gravitational effect predicted.

In June of 1976 a rocket probe carrying a hydrogen maser oscillator was fired to about a 10,000 mile altitude by NASA. This project, planned and carried out by VESSOT and LEVINE of the Smithsonian Astrophysical Observatory with NASA support [3.32], represented the culmination of several years of intensive effort to develop the packaged maser and its associated instrumentation and communication gear. The frequency difference between the probe clock and one on the ground, after elimination of Doppler effects due to the changing length of the communication link, agreed with predictions from the principle of equivalence to about $2 \times 10^{-4}$ for a major segment of the approximately 100 min useful part of the trajectory.

### 3.4.1 Future Possibilities

Further study of the effect of gravity on radiation as an application of the Mössbauer effect will need to introduce important improvements in resolution, if the evidence gained from the rocket probe experiment is to be surpassed. One technique that should make possible an experiment of precision similar to or even greater than that of the rocket probe would be the employment of a "light pipe" to reduce the losses associated with increasing the path length. If intensity were not decreased with increased distance, the statistical signal-to-noise ratio would improve linearly with an increase of the height h. A duct for $\gamma$ rays, based on total external reflection, can be made using a glass or metal pipe [3.33]. Those $\gamma$ rays striking the walls nearer to grazing incidence than the critical angle for total external reflection will be reflected successively from side to side as the ray moves along the pipe. The critical angle for total external reflection is proportional to the wavelength and to the electronic density. Experiments have demonstrated reflection with about 10% loss in ordinary glass laboratory tubing. It seems likely that glass of higher surface quality, such as plate or float glass, will be considerably better. The construction of a guide from flat plates probably coated with nickel inside a vacuum pipe seems a reasonable approach. The behavior should be similar in many respects to the ducts now employed at several reactor facilities to guide cold neutrons. The wavelengths, absorptions, and refractive index of materials for cold neutrons and for low energy $\gamma$ and X rays are very similar.

For an application employing such a guide, the 6.25 keV $\gamma$ ray of $^{181}Ta$ is particularly well suited. The critical angle is about $5 \times 10^{-3}$ radians for glass and $9 \times 10^{-3}$ radians for nickel. A solid angle of capture of $\gamma$ rays from a source would therefore be about that of the absorber seen from the source in the $^{57}Fe$ experiment of POUND and SNIDER. Even with the limited resolution so far achieved [3.34] (about 0.2 for the depth F and $V_H$ of 35 $\mu$m/s which is almost a factor of ten poorer than if the linewidth were down to the limit set by the lifetime of the level) use of

$^{181}$Ta with a $\gamma$-ray guide could provide a sensitivity competitive with that of the rocket probe. The statistical uncertainty derived from (3.2) using reasonable system parameters could be as small as $10^{-4}$ of the gravitational effect in a 250 m path. To carry out such a project, however, involves much development and expense. It is unlikely it will be carried out unless the full resolution available in principle from $^{181}$Ta is more nearly realized. If that were achieved, such an experiment could improve the resolution by a factor of ten or more beyond that of the rocked launched hydrogen maser. In such an event, investment in such a project would be quite attractive.

There are, of course, possibilities that Mössbauer resonances of extraordinarily greater resolution will be realized. It is estimated that an experiment competitive with the 1% precision of the $^{57}$Fe experiment could now be carried out with a $^{67}$Zn system using only about 1 m of height and a single $^{67}$Ga source approaching 1 Ci initially. The half-life of the $^{67}$Ga parent is only 76 hours, so such an experiment would be performed in about one week of data taking. It would probably be difficult but not impossible to avoid deleterious relative vibrations of the source and the absorber in a cryogenic system even as small as a meter. The most extravagant component of such a system would be enriched absorber which, at current prices, would be valued at about \$3000/cm$^2$.

Finally, of course, the realization of the resolution inherent in such long-lived isomers as $^{109}$Ag would open a new field of exploration, but such a development does not seem imminent.

## References

3.1 R.L. Mössbauer: Z. Phys. *151*, 124 (1958); Naturwissenschaften *45*, 538 (1958); Z. Naturforsch. *14a*, 211 (1959)
3.2 A. Einstein: Jahrb. Radioakt. Elektron. *4*, 411 (1907); Ann. Phys. (Leipzig) *35*, 898 (1911)
3.3 R.V. Pound, G.A. Rebka, Jr.: Phys. Rev. Lett. *3*, 439 (1959)
3.4 S. Devons, A.J.F. Boyle, D.St.P. Bundury: private correspondence from S. Devons (November 17, 1959)
3.5 T.E. Cranshaw as reported by J.P. Schiffer and W. Marshall: Phys. Rev. Lett. *3*, 556 (1959)
3.6 R.H. Dicke: private correspondence (November 12, 1950)
3.7 I.Ya. Barit, M.I. Podgoretskii, F.L. Shapiro: Zh. Eksp. Teor. Fiz. *38*, 301 (1960); [Engl. Transl.: Sov. Phys.-JETP *11*, 218 (1960)]
3.8 Koichi Shimoda: private correspondence (December 14, 1959)
3.9 W.S. Adams: Proc. Nat. Acad. (USA) *11*, 382 (1925)
3.10 G.E. St. John: Astrophys. J. *67*, 195 (1928)
3.11 M.G. Adam: Mon. Not. Roy. Astron. Soc. *108*, 446 (1948)
3.12 E. Finlay-Freundlich: Ann. Phys. (Paris) *2*, 765 (1957)
3.13 J.E. Blamont, F. Roddier: Phys. Rev. Lett. *7*, 437 (1961)
3.14 J. Brault: Bull. Am. Phys. Soc. *8*, 28 (1963)
3.15 J.L. Snider: Phys. Rev. Lett. *28*, 853 (1972)

3.16 S.F. Singer: Phys. Rev. *104*, 11 (1956)
3.17 D. Kleppner, R.F.C. Vessot, N.F. Ramsey: Astrophys. Space Sci. *6*, 13 (1970)
3.18 W. Wildner, U. Gonser: J. Phys. (Paris) *40*, C2-47 (1979)
3.19 A.H. Muir, F. Boehm: Phys. Rev. *122*, 1564 (1961)
3.20 R.S. Raghavan, L. Pfeiffer: Phys. Rev. Lett. *32*, 512 (1974)
3.21 D.G. Douglas: Can. J. Phys. *47*, 1815 (1969); R.S. Raghavan: Z. Phys. *243*, 441 (1971)
3.22 L. Pfeiffer: AIP Conf. Proc. *38*, 23 (1977)
3.23 R.V. Pound, G.A. Rebka, Jr.: Phys. Rev. Lett. *4*, 337 (1960)
3.24 R.V. Pound, J.L. Snider: Phys. Rev. Lett. *13*, 539 (1964); Phys. Rev. *140*, B788 (1965)
3.25 T.E. Cranshaw, J.P. Schiffer, A.B. Whitehead: Phys. Rev. Lett. *4*, 163 (1960); Proc. Phys. Soc. (London) *84*, 245 (1964)
3.26 R.V. Pound, G.A. Rebka, Jr.: Phys. Rev. Lett. *4*, 274 (1960)
3.27 B.D. Josephson: Phys. Rev. Lett. *4*, 341 (1960)
3.28 R.V. Pound, G.B. Benedek, R. Drever: Phys. Rev. Lett. *7*, 405 (1961)
3.29 J.C. Hafele, R. Keating: Science *177*, 166 (1972)
3.30 L. Briatore, S. Leschiutta: Nuovo Cimento *37B*, 219 (1977)
3.31 C. Alley et al.: presented orally at Symp. on Experimental Gravitation, Pavia, Sept.17-20, 1976
3.32 R.F.C. Vessot, M.W. Levine: *Gravitatione Sperimentale* (Accad. Naz. dei Lincei, Roma 1977) p.371; Gen. Relativ. Gravit. *10*, 181 (1979)
3.33 W.T. Vetterling, R.V. Pound: J. Opt. Soc. Am. *66*, 1048 (1976); R.V. Pound, W.T. Vetterling: J. Phys. (Paris) *40*, C2-3 (1979)
3.34 G. Kaindl, D. Salomon, G. Wortmann: Phys. Rev. *B8*, 1912 (1973)

# 4. Trends in the Development of the Gamma Laser

V. I. Goldanskii, R. N. Kuzmin, and V. A. Namiot

**With 4 Figures**

The history of lasers dates as far back as 1917 when Albert Einstein showed that the probability of a transition in a two-level system is characterized by the sum of two terms, one of which corresponds to spontaneous and the other to induced transitions. Almost a quarter of a century later in 1939, Fabrikant suggested for the first time the possibility of creating an inverted population of levels in an optical medium in spite of the Boltzmann distribution law so that the particle population $N_2$ of the upper (second) level exceeds the population $N_1$ of the lower energy level.

After Basov, Prokhorov and Townes invented, in the fifties, the maser principle of amplification and coherent generation in the radio-frequency (RF) region, other frequency ranges were investigates as well. In 1958, Shawlow and Townes showed that it was possible to construct an optical laser, two years later Maiman created a pulsed ruby laser, and in 1961 a continuous-wave helium-neon laser appeared. Ever since, lasers with increasingly harder radiation and higher power output have been produced.

From the standpoint of certain possible future applications, it is essential that the emission wavelengths of the lasers discussed in this paper become commensurate with molecular sizes. We are speaking of values in the neighborhood of $\lambda \lesssim 1$-10 Å, or at least a hundred times shorter than wavelengths of the vacuum ultraviolet (VUV). There are several possibilities for constructing lasers operating at wavelengths much shorter than the visible region. In general, we can speak of three major types of possible coherent sources in this wavelength range:

1) gamma(-ray) lasers in which use is made of transitions between the energy levels of an atomic nucleus;
2) X-ray lasers using transitions between the energy levels of atoms, with participation of deep electronic shells; and
3) electrodynamic lasers operating in the same general energy range as 1) and 2), including lasers based on free-electron beams in periodic magnetic fields [4.1-3], lasers based on the inverse Compton effect [4.4,5], and others.

This paper is concerned with the first type – gamma lasers. The other two are mentioned but briefly.

The basic drawback of X-ray lasers, as opposed to gamma lasers, is the broad level width of about 1 eV, which corresponds to extremely short lifetimes of about

$10^{-15}$ s. Therefore, for X-ray lasers to operate at wavelengths of 1 to 10 Å, all other things being equal, a pumping power exceeding that necessary for a gamma laser by at least a million-fold is required.

It has been estimated [4.6] that the specific excitation power Q is proportional to $\lambda^{-4}$ with $Q \gtrsim 10^{17}$ W/cm$^3$ at $\lambda \sim 1$ Å and $Q \sim 10^{13}$ W/cm$^3$ at $\sim 10$ Å. None of the available pumping techniques are effective enough to ensure such high specific power values.

In this respect, at wavelengths shorter than 10 Å where the transition energy becomes measurable in keV, gamma lasers are more promising than X-ray lasers. However, in order to make more realistic comparative predictions for various short-wave lasers, one should take many other factors into consideration apart from the pumping power, which is precisely what we will do here in discussing gamma lasers.

## 4.1 Formulation of the Problem

The possibility of creating a gamma laser based on the use of isomeric nuclear transitions [4.7-14] began to be discussed immediately after MÖSSBAUER discovered the effect now known by his name [4.15,16].

It soon became clear, however, that the problem of developing a gamma laser based on long-lived nuclear isomers could not be attacked frontally because of formidable difficulties and seemingly insurmountable obstacles; thus interest in gamma lasers wilted. Attention was renewed in the seventies coinciding with ideas for artificially suppressing inhomogeneous broadening of Mössbauer lines [4.17,18], for pulsed pumping of gamma laser [4.19], and for laser separation of nuclear isomers [4.20]. Of particular interest is Fig.4.1 which shows how publishing activity in the gamma-laser field has varied with time. Note two peaks: a peak of short duration in the early sixties and a longer one in the early seventies, the latter trend being now in a decline. All the works represented graphically in Fig.4.1 are theoretical papers

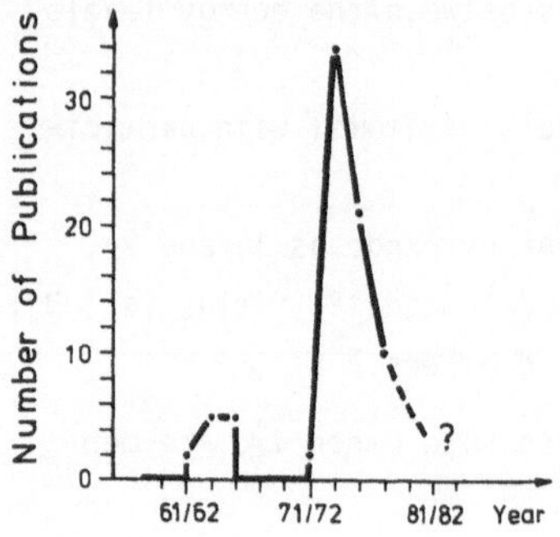


Fig. 4.1. Publishing activity in the gamma-laser field

and reviews, consequently experimental verification of the proposed ideas is now required as well as the generation of new ideas.

The problem of creating gamma lasers is covered at various lengths in several reviews [4.21-27]. We believe that our review will be of interest as well.

To provide for stimulated emission in a conventional two-level system, one must ensure populations per unit volume $N_1$ and $N_2$ of the lower 1) and upper 2) level such that the gain factor at the line center,

$$K = 2\pi\lambda^2 \frac{\Gamma_0}{\Gamma}\left(N_2 - \frac{g_2}{g_1} N_1\right) \quad , \tag{4.1}$$

exceeds the attenuation factor $\beta$, which for energies up to 100 keV is determined primarily by photoelectric absorption. Here, $\text{ⲭ} = 2\pi\lambda$ is the wavelength; $\Gamma$ is the effective line width which is larger, generally speaking, than the natural width $\Gamma_0 = \hbar/\tau$ because of various broadening mechanisms; $\tau$ is the lifetime for the spontaneous transition $2\rightarrow 1$; and $g_1$ and $g_2$ are the statistical weights of the operating levels.

The specific energy which must be introduced into the medium in order to achieve a given value for $N_2$ is

$$J = \hbar\omega N_2/\eta \qquad [\mathrm{J/cm^3}] \quad , \tag{4.2}$$

which requires for the specific pumping power

$$Q = J/\tau \qquad [\mathrm{W/cm^3}] \tag{4.3}$$

where $\eta$ is the efficiency, and $\tau$ is the effective excited-state lifetime. The threshold values of J and Q are determined from the condition $K = \beta$. The operating levels of X-ray lasers, or rasers, can be either intraatomic electronic levels or the higher levels of multiply-charged ions. In the former case, when an electron is selectively removed from the K-shell for example, an excited state with a K-hole occurs, and this could be the upper operating level which must be inversely populated. In the latter case, the function of the active medium is performed by a plasma with a non-equilibrium spectrum of free electrons. For the upper operating level to be filled, the participation of recombination electrons cascading down through the ionic levels from the continuum is essential. This is the principle of the plasma laser [4.28].

Since the photoabsorption cross section grows rapidly as the principal quantum number decreases, the selective removal of inner electrons may, in fact, be carried out by filtered bremsstrahlung or characteristic X-radiation at an appropriate wavelength from another element or from laser plasma. Equation (4.2) is not too critical with respect to the pumping source, e.g., for $\lambda \sim 1$ Å, $N_2 \sim 10^{22}$ cm$^{-3}$ and $\eta \sim 1$, we find $J \sim 10^5$ W/cm$^3$. Thus, it would be quite sufficient to have a source delivering an energy of about 10 J into a volume of $10^{-2} \times 10^{-2} \times 1$ cm. If, however, the upper operating state is used directly as an accumulator of X-ray energy, i.e., $\tau \sim \tau_{X_1}$, then (4.3) will impose highly stringent requirements on the rate of energy input from the source ($\sim 10^{15}$-$10^{21}$ W/cm$^3$ because in the X-ray range, $\tau_X \sim 10^{-10}$-$10^{-16}$ s). Conventional X-ray sources cannot meet such requirements.

Gamma lasers, or gasers, must operate at nuclear gamma transitions for which $\tau_\gamma$ greatly exceeds the lifetime $\tau_X$ of an X-ray transition. Therefore according to (4.3), producing the critical inversion in a gaser requires a much lower specific pumping power than for a raser. The expression for gain factor for Mössbauer quanta differs from (4.1) only by the factor $f(1+\alpha)$, where f is the Mössbauer-effect probability and $\alpha$ is the internal conversion coefficient, which is the ratio of the number of conversion electrons to the number of gamma quanta emitted during deactivation of the excited level ($\alpha=\eta/1-\eta$, where $\eta$ is the probability of deactivation of the excited level by emission of a conversion electron).

## 4.2 General Considerations

We shall begin by considering whether the conditions under which stimulated emission of gamma quanta occurs are compatible with the physical properties of real substances

As is known, a necessary (but not always sufficient) condition for induced emission is satisfaction of the inequality $n^*\sigma_\gamma > \sum_i N_i\sigma_{i(abs)}$, where $n^*$ is the concentration of working nuclei (i.e., of the nuclei at the upper level) present in the system, $\sigma_\gamma$ is the cross section for stimulated gamma-emission, $\sigma_{i(abs)}$ is the total cross section for all nuclear and electronic processes resulting in the absorption of gamma quanta by the atoms of the $i^{th}$ species.

In a first approximation, the variation in $\sigma_{i(abs)}$ can be ignored as one passes from visible light to gamma emission. As far as the value of $\sigma_\gamma$ is concerned, this resonance cross section varies as the square of the wavelength ($\lambda\!\!\!^-{}^2$), so that at energies of 1-100 keV it is lower by several orders of magnitude than in the case of visible light. Therefore, even without taking into account the broadening of the excited levels which is more pronounced for one of these levels with its longer lifetime $\tau_0$ and greater transition energy $E_0$, gamma lasers require a concentration of emitters greater by several orders of magnitude than in the case of optical lasers.

It has been shown [4.19] that the critical concentration of active nuclei of gamma emitters is

$$n_c^* = \left(\frac{E_0}{\pi\hbar c}\right)^2 \frac{\Gamma}{\Gamma_0} \frac{1+\alpha}{f\xi} \frac{1}{\ell(E_0)} \quad [\mathrm{cm}^{-3}] \quad , \tag{4.4}$$

where $\Gamma_0 = \hbar/\tau_0$ is the natural width and $\Gamma$ is the actual width of the emitting level, f is the fraction of deactivation processes for which the level width does not exceed $\Gamma_0$ (e.g., f is the Mössbauer-effect probability if by $\Gamma$ is meant the actual level width in the absence of nuclear recoil and Doppler broadening), and $\ell(E_0)$ is the mean free path of resonance quanta in the substance [at $E_0$=10-100 keV, $\ell(E_0) \leq$ 1 cm in the condensed phase]. $\xi$ stands for the so-called isomeric ratio $\xi = (N_2-N_1)/N_2+N_1) = 2\varkappa - 1$, where $\varkappa = N_2/(N_2+N_1)$ is the relative population of the upper level of the working nuclei of a gamma laser or the probability of populating this particular operating level in the course of pumping (for example, the probability that a Mössbauer nucleus formed by neutron capture will cascade to the upper level of the Mössbauer transition before decaying to the ground state). A necessary condition for stimulated emission to occur is $\varkappa > 1/2$.

Expressing $E_0$ in keV, one can write (4.4) as

$$n_c^* \approx 2.6 \times 10^{14} \frac{\Gamma}{\Gamma_0} \left[E_0^2 \frac{1+\alpha}{f\xi} \frac{1}{\ell(E_0)}\right] \quad . \tag{4.5}$$

For $E_0 \sim 10$ and $\alpha \sim 100$ or $E_0 \sim 100$ and $\alpha \sim 1$, then even with $f \sim 1$ and $\xi \sim 1$ we have $n_c^* \approx (10^{18}\text{-}10^{19})\Gamma/\Gamma_0$ cm$^{-3}$. This means that the broadening of the resonance lines used in gamma lasers must never exceed 3 to 4 orders of magnitude. Otherwise, the critical concentration may be greater than the total density of the substance [additional compression to superhigh densities does not help, since the mean free path $\ell(E_0)$ decreases inversely with density].

The above condition can be satisfied by Mössbauer spectra, although this requires that the working substance of the gamma laser be in a solid and a sufficiently cold state. Another extreme case in which relatively favorable $\Gamma/\Gamma_0$ ratios can be maintained is a non-Mössbauer resonance in a gas where Doppler broadening is present, but small in comparison to the natural line width. Under such conditions, $\Gamma/\Gamma_0 \approx 7 \times 10^{13} E_0 \tau_0 \sqrt{kT/A}$, where $E_0$ is expressed in keV, kT in eV, $\tau_0$ in seconds, and A is the mass number of the working nuclei. It can easily be seen that to provide for stimulated gamma emission without making use of the Mössbauer effect would require extremely low values of $\tau_0$ commensurate with X-ray transition times or slightly longer. Therefore, the non-Mössbauer variant of stimulated gamma emission necessitates much greater pumping power ratings to say nothing of a number of other serious difficulties [4.29,30]. This is why we shall restrict ourselves in what follows to the Mössbauer gamma laser.

Many years of studies of the Mössbauer effect have shown that unbroadened lines ($\Gamma \sim \Gamma_0$) are observed only in the case of sufficiently short-lived levels ($\tau_0 < \tau_{0max} \sim 10^{-6}\text{-}10^{-5}$ s), whereas for $\tau_0 > \tau_{0max}$ the line width increases as $\Gamma \sim \Gamma_0 \tau_0/\tau_{0max}$. Responsible for such an increase are various kinds of inhomogeneous line broadening which will be discussed in greater detail.

The foregoing suggests two possible ways of producing stimulated emission of Mössbauer gamma radiation: 1) Use of a nuclear species having an excited level with a sufficiently large natural width, which means a short-lived isomer with $\tau_0 \leq (10^3\text{-}10^4)\tau_{0max}$ or $\tau_0 \lesssim 10^{-2}$ s. This variant, first proposed in [4.19] and then elaborated in [4.31-33], calls for pulsed pumping of the gamma laser. 2) Use of various

Table 4.1. Characteristics of lasers using short- and long-lived Mössbauer nuclei

| Characteristic | Laser using short-lived levels | Laser using long-lived levels |
|---|---|---|
| 1. Level lifetime $\tau_0$ | $\lesssim 10^{-2}$ s | $\lesssim 1000$ s |
| 2. Preparation method | Exposure of the working substance to powerful pumping pulses, for example | Preparation of crystals with a high content of the long-lived working isomer |
| 3. Starting method | Pumping pulse | Application of external RF fields to suppress inhomogeneous line broadening |
| 4. Composition of working substance | Light matrix (e.g., beryllium) with small added amounts of nuclei which turn into working nuclei during the pumping pulse | Crystal containing working radioactive nuclei, e.g., in the form of whiskers at as high a concentration as possible |
| 5. Shape of the working substance | Acicular | Acicular |
| 6. Possibility of laser separation of isomers | Highly desirable for separating out the excited working nuclei after the pumping pulse. Limits the lower value $\tau_0$ | It is possible to prepare a starting material for subsequent crystal growth |
| 7. Basic operating principles | Ratio $\Gamma/\Gamma_0$ is not large to begin with owing to use of sufficiently large $\Gamma_0$ (i.e., small $\tau_0$) | Decrease in $\Gamma/\Gamma_0$ by suppression of inhomogeneous line broadening |

(primarily radio-spectroscopic) techniques to suppress inhomogeneous broadening of the resonance line. Taking $\Gamma \sim 10^{-15}$ eV (based on nuclear-magnetic-resonance data) as the minimum line width attainable with such techniques, the upper limit for the lifetime of excited working nuclei in a gamma laser turns out to be $\tau_0 \lesssim 10^3$ s. This variant, mentioned first in [4.7,9], laser revived by KHOKHLOV [4.17] and further elaborated in [4.18], will be referred to as the gamma laser using long-lived isomers. It does not require powerful pulse pumping but there are serious difficulties of which we shall discuss later.

A common feature of all variants of the Mössbauer gamma laser is the need to use crystalline solids with a concentration of radioactive isotopes which are either present or are created by a pumping pulse, and sufficient to provide stimulated gamma emission. The spontaneous emission by the excited nuclei can easily be amplified if a pencil beam of gamma rays is formed; therefore the working substance of the laser should preferably be needle shaped (acicular).

This shape for the working substance also has the advantage of minimizing heating of the subtance during the pumping pulse. An example is the case of pumping by radiative neutron capture where a significant fraction of not only the capture-gamma quanta themselves, but also the Compton electrons ejected by them, would escape through the sides of the needle without depositing all of their energy within the working substance.

The needlelike shape is also advantageous for lasers using long-lived isomers: it is known that "whiskers" of a sufficient size can be formed in a few seconds in a gas flow. Table 4.1 summarizes the characteristic features of the two types of gamma lasers using either short- or long-lived isomers.

## 4.3 Gamma-Laser Using Long-Lived Isomers

The early works dealing with gamma lasers [4.7-14] suggested the use of long-lived isomers with lifetimes of several days to a month. According to the authors, it would be sufficient to simply accumulate a certain number of excited nuclei and make a crystal out of them; the resulting gamma laser would then work. However, as was found out later, this is not possible because of the broadening of the gamma-quantum line. The ratio $\Gamma/\Gamma_0$ turns out to be about $10^6$, even for isomers with $\tau_0 \sim 1$ s, whereas for isomers with $\tau_0 \sim 10^5$-$10^6$ s the ratio varies from $10^{11}$-$10^{12}$. These values are far in excess of the maximum value, $\Gamma/\Gamma_0 \sim 10^3$, discussed earlier.

Among the decisive factors determining the practical upper limit for $\tau_0$ are the various sources of inhomogeneous broadening of the Mössbauer resonance line. The initiators of such broadenings include: 1) isomeric shifts, 2) quadrupole interactions, 3) magnetic hyperfine interactions, 4) magnetic dipole-dipole interactions between nuclei, and 6) gravitational shifts. We shall not discuss broadening due to magnetic interactions at this point because one can, in principle, select a diamagnetic substance and try to minimize the dipole-dipole interaction between nuclei using conventional nuclear-magnetic-resonance (NMR) techniques (see below).

The main contributions to inhomogeneous broadening often come from the electromagnetic interactions 1) and 2), but primarily from 1), the isomeric shift. The isomeric shift is inevitably present in the Mössbauer effect, since a Mössbauer transition involves two nuclear levels with different charge radii R. This is different from the situation in nuclear magnetic resonance where the transitions occur between the hyperfine sublevels of the same nuclear level. Inhomogeneities which produce only a distribution of shifts of the hyperfine structure "comb" center of the nuclear level have no significant effect on the spacings of the hyperfine levels nor on the NMR spectrum.

It is well known that in the case of a fixed transition, the isomeric shift depends only on the density of electrons at the nucleus $\delta \sim |\psi(0)|^2$. In a first approximation, $|\psi(0)|^2 \sim 1/V_0$ ($V_0$ is the unit cell volume), and a local change in the density of the substance is followed by a shift of the nuclear line

$$\Delta E_i = (\Delta E_i)_0 \left|\frac{\Delta V_0}{V_0}\right| \quad . \tag{4.6}$$

Analysis of a wealth of experimental results and theoretical estimates gives $(\Delta E_i)_0 \sim 10^{-7}$ eV.

A similar proportionality to the value of $|\Delta V_0/V_0|$ is also characteristic of the variation in the quadrupole splitting $\Delta E_Q$ of a Mössbauer line. This splitting is caused by the interaction of electric quadrupole moments of each nucleus with the gradient of the electric field created by the electronic shells of the same atom and by neighboring ions. The proportionality factor derived from experiments and their theoretical evaluations gives $(\Delta E_Q)_0 \sim 10^{-7}$ eV.

Another factor leading to an inhomogeneous variation in the quadrupole interaction is local shear strain. The latter may be characterized by a variation in the ratio of the longitudinal to the lateral dimensions of the unit cell, i.e., c/a. As a result, we have for isotropic and shear strains, respectively,

$$\Delta E_Q \sim 10^{-7}\left|\frac{\Delta V}{V}\right| \text{ eV} \quad , \quad \Delta E_Q \sim 10^{-7}\left|\frac{\Delta(c/a)}{c/a}\right| \text{ eV} \quad . \tag{4.7}$$

Consider now the contribution of point defects. It is known that at great distances from defects

$$\left|\frac{\Delta V_0}{V_0}\right| \approx \beta\left|\frac{\Delta V_0}{V_0}\right|_0 \left(\frac{a}{r}\right)^3 \quad , \tag{4.8}$$

where a is the interatomic distance, r is the distance to the point defects, and the numerical coefficient $\beta \sim 0.1$. Let us assume a value of 0.1 for $|\Delta V_0/V_0|_0$. In anisotropic crystals the same relation as in (4.8) exists for $|\Delta(c/a)/(c/a)|_0$ as well and, what is significant, even in crystals with cubic symmetry. As can be seen for (4.6-8), the line broadening due to point defects is already approaching the natural width of the line when the relative point defect concentration has the critical value $c^* \sim 10^9\Gamma_0$ (in this equation and in what follows $\Gamma_0$ must be expressed in eV). Hence, the critical concentration is $c^* \sim 10^{-6}$ for an excited-state lifetime $\tau_0 \sim 1$ s and $c^* \sim 10^{-10}$ at $\tau_0 \sim 10^4$ s. These values already attest to the difficulty of using long-lived isomers in sources of coherent gamma radiation.

In this respect, dislocations are at least as troublesome. For edge and screw dislocations we have, respectively,

$$\left|\frac{\Delta V_0}{V_0}\right| \sim 0.1\,\frac{a}{r} \quad , \quad \left|\frac{\Delta(c/a)}{c/a}\right| \sim 0.1\,\frac{a}{r} \quad . \tag{4.9}$$

If we denote, by $\eta$ (expressed in $cm^2$) the critical dislocation density corresponding to $\Delta\Gamma \approx \Gamma_0$, then we have from (4.6,7,9)

$$\eta^* \sim 10^{31}\Gamma_0^2 \quad . \tag{4.10}$$

Hence, even for the relatively short lifetime $\tau_0 \sim 1$ s there corresponds a low critical dislocation concentration $\eta^* \sim 10\ cm^{-2}$, while at lifetimes of $\tau_0 > 10$ s even one dislocation per $cm^2$ cannot be tolerated. Making such dislocation-free crystals or even crystals with a small number of dislocations is an extremely difficult and time-consuming task.

Inhomogeneous shifts of lines in samples of finite dimensions are also caused by surface effects. As can be inferred from available theoretical estimates, the relative variation $\varkappa_m$ in interatomic distances diminishes with increasing distance inward from the surface to an extent not greater than $\varkappa_m = \varkappa_0/m^3$, where m is the number of atomic planes between the given point and the surface. Consequently, we have the following values for the critical number $m^*$ of atomic planes inward from the surface at which the critical line shift $\Delta\Gamma \approx \Gamma$ occurs (for $\varkappa_0=0.03$): $m^* \approx 200$ for $\tau_0 \sim 1$ s and $m^* \approx 4000$ for $\tau_0 \sim 10^4$ s. It should be remembered that the above values of $m^*$ are understated, primarily because of the low value of $\varkappa_0$ which is accepted here.

Nevertheless, the use of thin layers may be inevitable because of the short time available for preparing the emitters, or indispensable in order to minimize heating and to eliminate inhomogeneous gravitational shifts. The fractional shift in energy with vertical height is $\delta_{grav} = 10^{-18}$ $cm^{-2}$. Thus the line broadening would be $10^{-14}$-$10^{-13}$ eV/cm for E = 10-100 keV. A broadening of $\Delta|\varkappa|_0$ occurs for $\tau_0 \approx 1$ s when the vertical thickness is 0.1-1 mm and for $\tau_0 \sim 10^4$ s when the vertical thickness is 0.01-0.1 microns.

As for the temperature-induced red shift and Mössbauer line broadening, they can be avoided, in principle, if one operates at the lowest possible temperatures. Otherwise, for example at room temperature, limitation of inhomogeneous broadening to $\Delta\Gamma = \Gamma_0$ for $\tau_0 \sim 1$ s and $E_0 = 10$-100 keV, would require a temperature range $\Delta T \sim 10^{-4}$-$10^{-5}$ K. For $\tau_0 \sim 10^4$ s, a temperature range of only $\Delta T \sim 10^{-8}$-$10^{-9}$ K would result in a broadening $\Delta\Gamma = \Gamma_0$. At the same time, as was shown by KAGAN [4.34] even for $\Delta T = 0$ K, the homogeneous Mössbauer line broadening at room temperature would be $10^{-15}$-$10^{-13}$ eV for $E_0 = 10$-100 keV.

As has been mentioned above, magnetic hyperfine and dipole-dipole interactions occur in NMR as well. There are NMR techniques available which use specially selected applied RF fields to suppress the broadening due to these mechanisms [4.18,35].

The physical principle of suppression is based on the fact that nuclear moments can be turned in space by specially selected RF fields. A short train of RF pulses, called $\pi/2$ pulses, may be applied to the sample in such a way as to rotate the nuclear moments from one to another of a specially selected set of directions. The train of $\pi/2$ pulses can be selected so as to average the magnetic dipole-dipole interaction to zero and thereby reduce the line width by several orders of magnitude. Although the RF field also splits the line into a number of satellites, the width of each satellite remains acceptably small. This kind of averaging may be thought of as a "jolting" of nuclear moments which makes them insensitive to electromagnetic inhomogeneities in the crystal, as will now be demonstrated.

The Hamiltonian of the dipole-dipole nuclear interaction takes the form

$$H \sim (\hat{I}\hat{I}'-3\hat{I}_z\hat{I}'_z) \quad . \tag{4.11}$$

The rotation of nuclei through $\pi/2$ by an RF pulse requires replacement of this Hamiltonian by

$$H \sim (\hat{I}\hat{I}'-3\hat{I}_x\hat{I}'_x) \quad . \tag{4.12}$$

After another rotation of the nuclear moment through $\pi/2$, it becomes

$$H \sim (\hat{I}\hat{I}' - 3\hat{I}_y\hat{I}'_y) \quad . \tag{4.13}$$

Addition of (4.11-14) gives $\bar{H} = 0$. That is, the interaction is averaged to zero and on the average produces no disturbance of the energy levels. This procedure of RF averaging eliminates the local effects of magnetic inhomogeneities: all the nuclei find themselves in similar conditions, and the emission lines narrow. For Mössbauer nuclei, in contrast to NMR, the RF field rotates the moment of nuclei in both the ground and excited states. The narrowing effect may reach several orders of magnitude. Analysis [4.36,37] shows that at the same time there is also a reduction in the broadening due to the interaction between the quadrupole nuclear moment and the electric field gradient.

The isomer shift associated with the change in the nuclear volume during the transition has no analogue in NMR. The techniques available in NMR, which are based (as has just been described) on averaging of the interaction to zero over the nuclear lifetime, might seem to be useless in this case. Nevertheless, RF suppression (or rather compensation) of broadening caused by inhomogeneous isomer shifts also turns out to be possible [4.38].

For a better understanding of the idea behind this, let us consider Fig.4.2a, which represents the hyperfine structure of the nuclear levels. The splitting of the levels is caused by the magnetic field of unpaired electron spins. Let us assume that the volume of a unit cell accommodating a nucleus has changed as a result of a disturbance, consequently the electron density at the nucleus has also changed. This change in the density can be expressed as $\delta|\psi|^2_\downarrow + \delta|\psi|^2_\uparrow$, where $|\psi|^2_\downarrow$ is the density of electrons with "downward" spins and $|\psi|^2_\uparrow$ is the density of electrons with "upward" spins. This change in density changes the isomeric shift (Fig.4.2b) and at the same time, brings about a change in Zeeman splitting which is proportional to $\delta|\psi|^2_\downarrow - \delta|\psi|^2_\uparrow$. For small variations in $\Delta V_0$, the change in the isomeric shift is proportional to the change in the magnetic hyperfine splitting, and since this splitting can be controlled by external RF fields, the proportionality factor may be varied within certain limits. It may be selected in such a way that any changes in the isomer shift would, for one of the lines of the hyperfine structure, be compensated exactly by a corresponding change in the amount of hyperfine splitting (Fig. 4.2a,b).

Thus it appears that by using external RF fields one can, in principle, suppress all the detrimental electromagnetic mechanisms of Mössbauer line broadening. However,

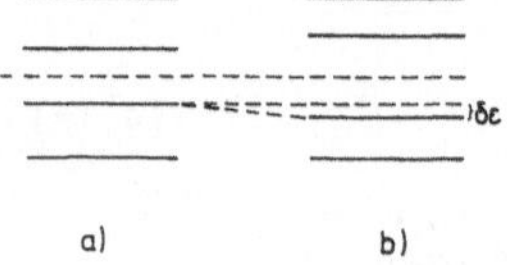


Fig. 4.2a,b. RF compensation of isomeric shifts. a) Hyperfine structure of the nuclear levels. b) Splitting of the levels δε by unpaired electron spins

considering the line narrowing experimentally attainable in NMR, a line with less than $\Gamma \sim 1$ Hz is very unlikely. As was previously shown, $\Gamma/\Gamma_0$ should not exceed $10^3$ or $10^4$. This means that the lifetime of long-lived isomers to be used in gamma lasers should not exceed the value already mentioned above: $\tau_0 \sim 1000$ s.

We should also like to point out that there are other ways of narrowing the line which have nothing to do with the effect of an RF field. First, there are other solutions to the problem of suppressing inhomogeneous isomeric broadening besides the compensation method that has just been described [4.33,36]. One may consider another possibility which arises when the isomer shift is varied by resonance repopulation of electronic levels during optical pumping. The sensitivity of level population to local changes in the frequency of the optical transition permits, in principle [4.39], simultaneous compensation throughout the crystal [4.33] for the inhomogeneous portion of the isomer shift and even contraction of the hyperfine structure into a single line [4.40]. Also possible is compensation for line broadening under specific conditions governing the Mössbauer effect. For example, experiments indicate that NMR lines for liquids are narrower by several orders of magnitude than for solids. This has to do with the averaging of disturbances due to motions of atoms. In Mössbauer spectroscopy, however, the motions of atoms would cause Doppler line broadening instead, so that it is impossible to apply such a line-narrowing method directly. Nevertheless, situations may arise in which averaging during motion will not be accompanied by Doppler broadening [4.41]. For example, atoms in the surface of a liquid can move freely along the surface and at the same time radiate without Doppler broadening in a direction normal to the surface [4.41]. Another example is atoms in a spatially periodic field on the surface of a crystal. If the spatial period of the field is a multiple of the wavelength of the emitted Mössbauer radiation, then for quanta emitted along the surface (in a direction coinciding with that of the wave vector of the field), emission without Doppler broadening is also possible [4.42]. The narrowing of a Mössbauer line is a prerequisite for operation of gamma lasers based on long-lived isomers; the only other approach is dynamic or static alignment (polarization) of the nuclear spins. Induced nuclear alignment leads to anisotropy of the emitted gamma radiation which provides a means whereby the beam of gamma quanta may be directed along the crystal's longitudinal axis [4.43].

## 4.4 Gamma-Laser With Pulsed Pumping

The main advantages of a gamma laser using short-lived nuclear isomers is that it becomes unnecessary to prevent Mössbauer line broadening by growing dislocation-free crystals, resorting to RF narrowing of lines, and so on. The $\Gamma/\Gamma_0$ ratio is quite acceptable even without such measures.

The basic principle of gamma laser operation with pulsed pumping is "instantaneous" (within $\tau_{min} \ll \tau_0$) pumping to a supercritical population of Mössbauer-excited nuclear states for which $\tau_0 \lesssim 10^{-2}$ s.

The possibility of providing adequate conditions for stimulated emission of Mössbauer gamma quanta ($E_\gamma \sim 5$-100 keV) was set forth for the first time in [4.19]. It proposed populating the working resonance level of nuclei with mass number (A+1) through radiative capture of neutrons by nuclei of a neighboring isotope of the same element:

$$ {}^{A}_{Z}X + n \rightarrow {}^{A+1}_{Z}X^{*} + \sum \gamma_{capture} $$
$$ \qquad \hookrightarrow {}^{A+1}_{Z}X + \gamma_{Mössbauer} \quad . $$

Pulsed pumping necessitates the use of neutron sources of extremely high power such as have not yet been developed. High neutron fluxes present a serious danger of excessive heating of a gamma laser's working substance while the required high concentration of short-lived Mössbauer emitters is being attained. Such heating may completely rule out the possibility of achieving the Mössbauer effect.

Thus, development of the pulse-pumped nuclear gamma laser imposes not only extremely stringent, but also conflicting requirements on the system parameters. The first step in the development of this laser must be a demonstration of the possibility of meeting these requirements. This must be followed by exploration of possibilities for improving pumping techniques and consideration of other design questions such as the integrated power, angular directivity, and emission kinetics of the nuclear laser.

The basic sources of heating for the working substance of a gamma laser during pumping by radiative neutron capture include hard (up to several MeV) capture-gamma quanta. These quanta transmit their energy mainly to Compton electrons, but also to recoil nuclei which acquire energy by elastic and inelastic scattering of neutrons.

Prevention of impermissible heating by recoiling nuclei requires slowing down of the pumping neutrons to energies not exceeding several tens of eV before they penetrate the working substance of the gamma laser. The slowing down of neutrons is also desirable from the viewpoint of increasing the cross section for their radiative capture. However, the required slowing down of the neutrons also imposes a lower limit on the deviation of the pumping pulse: $\tau_{pulse} \gtrsim 10^{-6}$ s.

To prevent undesirable heating due to the gamma quanta of radiative neutron capture, the working substance of the gamma laser must be needle shaped with a thickness $\ell$ less than the path length of the majority of Compton electrons: $\ell \sim 1$-10 μm.

A highly important factor which may materially facilitate the development of a pulse-pumped gamma laser is the use of light nuclei which absorb neutrons and gamma quanta only weakly as the basic material of the working substance. By this means one can substantially increase $\ell(E_0)$, the mean free path of resonance gamma quanta in the working substance, since $\ell(E_0)$ appears in the denominator of (4.5). This will increase the critical concentration of the working nuclei and at the same time effectively cool the medium by distributing the energy of radiative neutron capture throughout the matrix. The most suitable material for the matrix of a gamma laser seems to be beryllium. The optimum composition of the working substance within the matrix is determined by the relation $\sigma_M N_M \quad \sigma_W N_W$, where $N_M$ and $N_W$ are the concentrations of the matrix and working nuclei, respectively, and $\sigma_M$ and $\sigma_W$ are the cross sections for absorption of resonance gamma quanta by these nuclei.

If the mass number of the working nuclei is assumed to be $A \sim 150$, the optimum concentration of these nuclei in the beryllium matrix at $E_0 \sim 10$-100 keV must only be about $10^{-3}$-$10^{-4}$.

The isotope, $^{181}Ta$ ($E_0 \sim 6$ keV) pumped by radiative capture of neutrons in $^{180}Ta$, has been examined as a possible working nuclear species for a gamma laser [4.19]. In another work [4.44], there was discussed the possibility of using the 9.3-keV gamma transition in $^{83}Kr$, pumped either by the radiative neutron capture process $^{82}Kr(n,\gamma)^{83}Kr$, or by the photonuclear reaction $^{84}Kr(\gamma,n)^{83}Kr$, where a high-current electron accelerator would be used as the gamma source.

Thus it is possible, in principle, to reconcile the existing parameters of real materials with the requirements of an inverted population of a gamma laser while avoiding undesirable heating. Hence, the "frontal" approach involving pumping is possible, at least in principle, owing to the existence of a "window" of self-consistent physical parameters. In this case however, the required integrated neutron fluxes within a time interval $t \ll \tau_0$ are colossal: $I_t \sim 10^{19}$-$10^{20}$ cm$^{-2}$. Such intensities are attainable at present only in nuclear explosions (which, of course, would permit the pumping of a large number of needles at the same time).

Still, there is hope that gamma lasers can be made to operate with much lower neutron pumping fluxes. Indeed, the total number of active nuclei in a laser needle need be only $10^{13}$-$10^{14}$. This suggests the possibility of simultaneous excitation of Mössbauer levels (optimally by neutron capture) and transfer of the excited nuclei to a receiver. In this case, it would be possible to collect very short-lived nuclei and to separate the heating region where the excited levels are first populated from the laser region where the subsequent resonant gamma emission occurs.

All variants of the transplantation idea require collection of excited Mössbauer nuclei as they escape from the surface of a neutron-irradiated sample due to their recoil energy after radiative neutron capture. In the case of irradiation of highly dispersed (100 m$^2$/g) systems, even when only 1% of the irradiated volume (in the form of dust) is filled, one cubic centimeter contains $10^{19}$-$10^{20}$ nuclei ready to escape from the surface by recoil following neutron capture. For $\sigma_{n\gamma} = 10^{-21}$-$10^{-20}$ cm$^2$ and irradiation of 1-10 cm$^3$ of such dust, one can obtain about $10^{15}$ excited nuclei from as low an integrated flux as $I_t \sim 10^{15}$ cm$^{-2}$, this amount being sufficient for stimulated gamma emission. Fast collection of the escaping Mössbauer atoms is possible, for example, by transforming them into molecules of oxides (in an oxygen atmosphere) with subsequent entrainment in supersonic flows, or alternately, by ionization and subsequent focusing in an electric field. Hence, in these variants of the gamma laser, the working substance (needle, thin film, etc.) is formed from previously excited short-lived Mössbauer emitters, possibly interspersed with light nuclei. Note that attaining integrated fluxes of such sizes within time intervals $t \ll \tau_0$ of about $10^{-6}$ s are quite feasible [4.45], although the actual neutron sources are not yet available. The optimum solution may be use of special plasma sources which generate neutrons in thermonuclear, or fusion reactions, a possible example being "plasma focus". The problem of selecting an isotope which would be optimum in all respects remains open. At present, there is no single transition for which all the necessary characteristics are known, particularly the so-called isomeric ratio. In view of this, special investigations within the framework of nuclear spectroscopy are called for.

Another way to reduce the neutron flux required to pump a gamma laser is to use two-stage pumping, which was proposed in [4.46]. The basic principle is as follows. The capture of neutrons to form excited Mössbauer nuclei of mass (A+1) occurs in an intermediate solid target which contains the stable isotope of the same element having mass A. After a neutron has been absorbed and the "instantaneous" capture-gamma quanta have been emitted, the final nucleus is in the excited Mössbauer level of the isotope having mass A+1. This in turn passes to the ground state by emitting a resonance quantum. The quanta emitted in these processes then enter the working crystal which contains nuclei of the same Mössbauer isotope in the ground state, and there they are resonantly absorbed, provided the Mössbauer effect takes place during both emission and absorption. In this case if, as usually happens, the cross section for resonant capture of the gamma quanta in the working crystal ($\sigma_{\gamma\gamma}$) is much greater than that for radiative neutron capture in the intermediate target ($\sigma_{n\gamma}$), it may be expected that the absolute density of excited nuclei in the working crystal ($n_2^*$) will be much higher than in the intermediate target ($n_1^*$).

An optimum value of the ratio $n_2^*/n_1^*$ is 10-100 and requires Mössbauer excitation energies of 30-50 keV. In this variant of the gamma laser, the requirement of a favorable isomeric ratio following neutron capture,

$$n^*_{(A+1)}n_{(A+2)} > 1 \quad , \tag{4.14}$$

being extremely difficult to meet, becomes obsolete.

Two-stage pulsed neutron pumping will decrease the required neutron flux density by 1 to 2 orders of magnitude and in addition minimize heating of the working substance of the laser.

Both methods, two-stage pumping and transplantation, offer material advantages over the "frontal" pumping approach. Here, the region, where $(n,\gamma)$-pumping occurs and the elimination of heating is difficult, is spatially separated from the working substance of the gamma laser where the process of gamma generation occurs. In this variant, the virtual realization of gamma lasers based on short-lived isomers approach gamma lasers based on long-lived isomers.

All of the above-described methods of collecting nuclei fail to separate nuclei that have absorbed neutrons and decayed to the excited Mössbauer level from those that have absorbed neutrons and decayed directly to the ground level. In his work [4.20] LETOKHOV proposed an optical-laser method of rapidly separating excited from unexcited nuclei. The basic idea of the method is as follows. Nuclei in the excited state $(A+1)^*$ differ from those in the ground state (A+1) and all other nuclear isotopes in the target in that their electronic levels are somewhat shifted energetically. If an optical beam with a preselected radiation frequency is incident on such a mixture of isotopes, the frequency being such that optical quanta are resonantly absorbed only by the $(A+1)^*$ atoms, some atomic electrons of these atoms are transferred to higher excited levels.

Thus, atoms containing excited $(A+1)^*$ nuclei have been labelled, or excited, a second time. Let us denote such atoms by $(A+1)^{**}$. It is precisely these atoms that must be separated from the mixture. To this end, the evaporation products must be exposed to another powerful light source which easily ionizes $(A+1)^{**}$ atoms by ejecting the electrons which are now in excited levels, transforming these atoms into positively charge ions, $(A+1)^{*+}$. This corresponds to the atomic photoionization process. As a result, all the atoms in the mixture are neutral, except for the excited $(A+1)^{*+}$ ions where nuclei are in the excited state necessary for creating a gamma laser. Now, the $(A+1)^{*+}$ ions may be directed, with the aid of an external electric field, to a collector where a crystal is grown from them.

At present, lasers are used to separate some isotopes, especially those with high mass numbers. It can be pointed out that the optical-laser parameters necessary to implement this method of separating Mössbauer isomers have already been attained.

The above example, which includes a new application of optical lasers, is an additional demonstration of successful joint efforts by investigators working in different fields of science to solve the problem of gamma lasers.

Other methods or rapid separation of isomers, apart from those involving lasers, can be employed [4.47,48]. Consider now another possibility for minimizing the required neutron fluxes, this time by increasing the mean free path for nonresonant absorption of the resonance gamma quanta.

An increase in the mean free path of X rays moving in certain directions in a crystal was predicted by BORRMANN [4.49]. The Borrmann effect is the sharp decrease in the absorption coefficient of X or gamma rays, the so-called irregular transmission of X rays, which occurs during Bragg scattering. Closely similar to the Borrmann effect is its nuclear analogue, the suppression of inelastic channels of nuclear reactions [4.50] otherwise known as the Afanasyev-Kagan effect. This suppression effect exists because the interaction between nuclei and the electromagnetic (Mössbauer) radiation in a crystal is complex and under the same conditions as in the Borrmann effect. That is, when the Mössbauer resonance radiation passes through a crystal at the Bragg angle, the wave field is so distributed that the amplitudes for excitation nuclei sharply decrease. As a result, Mössbauer gamma quanta propagate in the crystal at the diffraction angle almost without absorption.

For a gamma laser it is obviously desirable to maintain the interaction of the gamma quanta with the nuclei while minimizing their interaction with electrons. In other words, it is desirable to eliminate the Afanasyev-Kagan suppression effect without diminishing the Borrmann effect. This can be achieved if high-multipolarity nuclear transitions are used as the working laser transitions [4.31,51-54]. The effect of irregular transmission of gamma quanta in an acicular crystal manifests itself in the same manner as in an ordinary crystal.

The diffraction of gamma quanta by a succession of many atomic Bragg planes prevents the transmitted beam from deviating from its initial direction, and therefore the transmitted gamma-quanta flux density remains the same over a long distance. Of course, the beam eventually spreads out due to the transverse diffusion of slowly changing field amplitudes. It may be shown that an incoming beam 0.1 mm wide becomes twice as wide at a distance of about 0.5 m [4.54].

## 4.5 On the Kinetics of Amplification of Emission in the Gamma-Laser

The kinetics of induced gamma emission started receiving widespread attention only recently. Investigations [4.8,55-66] show that the kinetic pattern describing the time dependence of the cross section for an induced process and of the waveform of an amplified signal may substantially alter the very concept of lasing threshold.

The use of computers to model the kinetics of stimulated gamma emission looks promising. The first examination of the kinetics of emission for a gamma laser based on long-lived isomers was made by CHIRIKOV [4.8]. He has shown that the characteristic gamma-wave evolution time under such conditions is much shorter than the radiative transition time $\tau_0$, and consequently the cross section $\sigma_i$ for induced emission is not constant but grows gradually during the course of this evolution, approaching the resonance cross section $\sigma_r$ at $t \to \tau_0$. This is due to the fact that although the gamma wave evolves over a short distance in the material (ca. 1 mm), the required time for this evolution is relatively long (ca. 1 s).

The fact that the cross section for induced emission of a gamma quantum changes during emission becomes clear from the following reasoning. If we had N excited nuclei with lifetime $\tau_0$ and if no gamma quanta were present at the initial moment, then after $t_1 \sim 1/N\tau_0$ another quantum would appear in the system. But the quantum "multiplication" time (provided absorption can be ignored) is $t_2 \sim 1/N\sigma_i c$, where $\sigma_i$ is the cross section for induced emission and c is the velocity of light.

However, if we mainly assume that $\sigma_i$ is the same as $\sigma_r$ and use $\sigma_r$ in the expression for $t_2$, we may arrive at the result that $\tau_2 << \tau_0$. This will bring us to an inconsistency, for in such a case, according to the uncertainty principle, the quantum energy cannot be determined with sufficient accuracy for $\sigma_i$ to be considered a resonance cross section. Hence, for such short time intervals one cannot adopt the equality $\sigma_i = \sigma_r$.

Dependent both on time and on the electromagnetic field is $\sigma_i$. The mathematically rigorous statement would be: $\sigma_i$ is a function of time and the vector potential of the electromagnetic field. Therefore, the following self-consistent problem in kinetics must be considered: $\sigma_i$ determines the growth of the photon avalanche, and at the same time the photon avalanche's electromagnetic field influences $\sigma_i$.

A rigorous approach, with both the electromagnetic field and the nuclear transitions treated quantum mechanically, turns out to be rather difficult. In [4.58-60], a semiclassical approximation is used, i.e., the electromagnetic field is considered to be classical while the nuclear transitions are considered to be quantum transitions. Spontaneous decay is assumed to be the initiating factor. In this case, the processes develop as follows. At the initial moment, the phases of the emitting dipoles (for simplicity we approximate the emitting nuclei by oscillating dipoles) do not correlate with one another. However, as more photons are exchanged, the correlation increases, and eventually they all emit with a single phase. The maximum correlation is attained after delay time

$$t_0 = \left(\frac{1}{\tau_c} - \frac{1}{T_2}\right)^{-1} \ln\left(\frac{T_1}{T_c}\right) , \tag{4.15}$$

where $T_c = T_2/\beta L$ and L is the length of coherent irradiation. At this instant, a short buildup of intensity is observed

$$I_{max} \simeq \frac{N}{4\tau_c}\left(1 - \frac{T_c}{T_2}\right)^2 , \tag{4.16}$$

where N is the number of particles in the system and $T_2$ the time of dephasing (time of phase relaxation). After this, light emission continues in a coherent manner, but faster than during spontaneous decay. Note that normal induced emission discontinues as soon as the population difference $[n_2 - n_1(g_1/g_2)]$ drops to zero. This process is referred to as superfluorescence as opposed to Dicke superemission and superluminescence (induced emission). Also note that faster light emission in superfluorescence does not provide any clues to the problems of line narrowing and pumping.

A radically new idea for a nonthreshold gamma laser (or, to be more precise, parametric amplifier) is advanced in [4.66-67]. The authors bring forward for the first time the possibility of parametric amplification of Mössbauer emission in a medium having an inverted population for a transition in the optical range.

Theoretical studies of the kinetics of emission of gamma lasers must proceed with due regard for all possible factors: spectral pumping density, temperature of the system, specific level lifetimes, line narrowing, crystal defects, and so on.

Without introducing specific parameters as well as characteristics of nuclei and crystals into gamma-laser models, analysis of the kinetics of light emission will be merely an exercise in abstraction.

## 4.6 Resonators

The basic components of any laser are an inverted medium, a pumping source, and finally, a resonator. By analogy with optics it is easy to define the purpose of the resonator of a gamma laser. The resonator must provide positive feedback, thereby enhancing the Q factor, improving monochromaticity, controlling the beam divergence, determining the modal structure of emission, and minimizing the pumping power. Beam reversals are quite easily accomplished in optics. Beams are reflected from mirrors and pass several times through an active medium without significant loss because of the superior quality of the mirror surfaces available for this region of the spectrum. In the case of hard ultraviolet, an adequate resonator becomes more difficult to make, while in the case of X and gamma radiation the difficulties are not yet completely known.

Most proposed resonators for $\lambda \sim 1$ Å have periodic structures. The first such proposal was made by RIVLIN [4.68]. His idea was based on the fact that in an ideal crystal it is easy to use Bragg diffraction to provide a closed beam path. Estimates [4.68,69] show that for the X- and gamma-ray wavelength range, the characteristics of resonators based on single crystals are as good as those of resonators for the optical range.

It is well known that for ideal single crystals where the Wulff-Bragg condition is met (i.e., $2d \sin\theta = n\lambda$, where $\theta$ is the diffraction angle and $n = 1,2,\dots$ is the order of diffraction) specular reflection is maintained. The higher n is, the greater $\theta$ can be. At $2\theta = 180^\circ$, $d = n\lambda/2$ and backward reflection is possible.

Hence, distributed feedback for X and gamma rays may be achieved by using Bragg diffraction after proper selection of both the crystal and the wavelength. In the case of X rays we are dealing with the characteristic emission excited by electron bombardment (pumping). In the device of Fig.4.3a [4.70], waves traverse $10^2$-$10^3$ cells within the characteristic time of X-ray fluorescence ($10^{-14}$-$10^{-15}$ s). In a periodic medium, photoabsorption may become very weak because of coherent wave superposition giving rise to nodes of the electric field intensity at the sites of the most absorbing atoms.

An alternative resonator was proposed, based on natural resonation in zeolite crystals [4.71]. These crystals are essentially porous aluminosilicate structures forming so-called molecular sieves with passages 3-10 Å wide interconnecting cavities up to 25 Å in diameter. Figure 4.3b shows a passage in a longitudinal section.

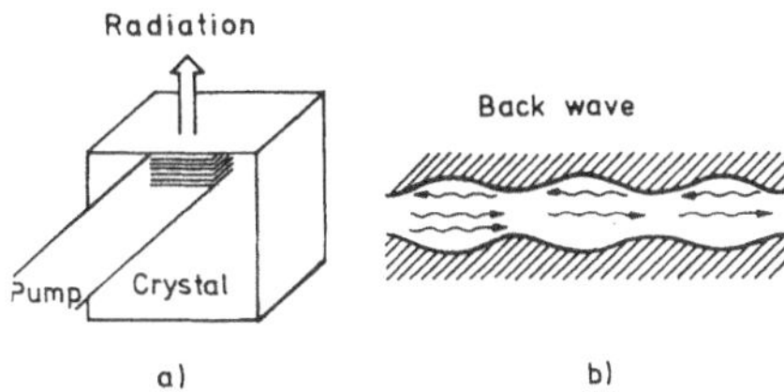


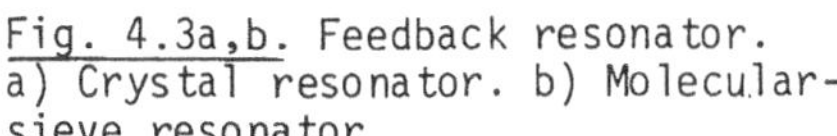

Fig. 4.3a,b. Feedback resonator. a) Crystal resonator. b) Molecular-sieve resonator

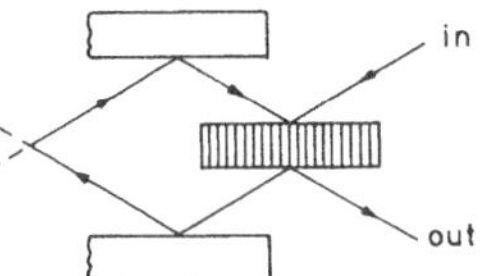


Fig. 4.4. Resonator crystal with excited nuclei

Here, distributed feedback is ensured by periodic variation of the passage cross section and diffraction scattering of quanta from their inner surfaces. In such a system, the pronounced distributed feedback necessary for self-excited vibrations can be achieved.

At this stage purely Bragg "circular" systems seem to offer the most realistic opportunities. Figure 4.4 represents one such system. It includes an active crystal in which the Borrmann effect occurs, provided the diffraction condition is met [4.69,72]. It is also known that the reflection of X or gamma rays may be highly efficient (the coefficient of reflection from an ideal single crystal may approach unity). There are absolutely reliable data attesting to 95% efficiency of reflection of a parallel Cu $K_\alpha$-radiation beam from the (220) Bragg's reflection planes of a perfect germanium crystal in the case where polarization is normal to the scattering plane [4.73].

From the practical standpoint, one must have a resonator tunable to any wavelength. COTTERILL [4.74] has proposed a tuned resonator meeting this requirement in which an even number of reflectors with parallel pairs of crystals is to be used. The number of pairs (N) is determined from the inequality $N \cdot 2\theta > 180^\circ$. The path symmetry is independent of $\theta$, and the resonator can be tuned to any wavelength.

We do not see any apparent physical reasons why high-Q resonators cannot be made for X and gamma rays. Technically, the task remains highly complicated although it can be resolved by state-of-the-art techniques of producing and processing crystals. As was pointed out in a review [4.75], an X-ray beam focuses itself in a ring resonator with attenuation ceasing after several reflections and line width also narrowing significantly.

Having much in common, resonators for gamma- and X-ray lasers will nevertheless be different. In resonators for gamma rays, in addition to the Borrmann effect [4.48], use should also be made of the inelastic channel suppression effect [4.48] and the effect of redistributing the nuclear reaction products in crystals [4.76].

## 4.7 Short Wave Radiation from Moving Charged Particles

The excitation of atoms and nuclei is known to occur during interactions of charged particles with a substance. The excitation of atoms by electrons results in bremsstrahlung and characteristic X-radiation spectra. Over the past ten to fifteen years various theories concerning coherent X radiation have appeared, based on the interaction between a beam of charge particles (electrons, protons, ions) and a substance. In most cases, the presence of a periodic medium, such as a single crystal, is required. In some cases the orientation of the medium with respect to the particle beam is of primary importance.

Let us first analyze works predicting the possibility of hard electromagnetic radiation of the transition or Vavilov-Čerenkov type in crystals. In his paper [4.77] RIVLIN hypothesized that if velocity v of an electron moving in a single crystal coincided in value and direction with that of one of the harmonics of the wave radiation field, the Vavilov-Čerenkov radiation in the X-ray frequency range may appear. In other works [4.78,79] he suggests forming pseudocrystalline discrete electron beams in which electrons are distributed in a periodic manner. The interaction between periodically ordered beams of charged particles and a single crystal may bring about selective excitation or ionization, including that of deep electronic levels. This effect could be used to produce a population inversion.

BELYAKOV [4.80] analyzed the possibility of producing coherent structural Vavilov-Čerenkov radiation in the X-ray frequency range during passage of a channelized charged particle through a crystal. In this case the emission of a photon by an individual crystal atom is induced by the Coulomb field of the channelized particle. A distinctly anisotropic coherent emission of gamma quanta is possible if this Coulombic process excites Mössbauer gamma resonances [4.81-83]. To obtain a pronounced cross section for coherent excitation, relativistic particles should be used, the most suitable for the purpose being fast electrons [4.84]. In some works [4.82,85] it is contended that fast electrons produce gamma radiation of the Čerenkov or transition type. As is pointed out in the latter work [4.85], a current of 10 A ($6.3\times10^{10}$ el/s) may yield only several tens of quanta per second. This is of interest, nevertheless, because of the pronounced directivity of such radiation. The main processes competing with Čerenkov radiation are bremsstrahlung and transition radiation, the latter having the same spectral composition as Čerenkov radiation.

It should be noted that incoherent Coulomb excitation is already used in the making of Mössbauer sources and is normally accomplished with the aid of nonrelativistic particles (protons, ions).

The authors of [4.86] have given consideration to the possibility of both direct and stepwise excitation of Mössbauer levels in the accelerated nuclei themselves (beams of heavy ions) by Coulomb excitation occurring as the nuclei come to a stop in the substance.

Table 4.2. Estimates of excitation probabilities for $^{57}Fe$ resulting from various types of bombardment

| Bombarding particles and nuclear reaction | Yield of excited nuclei | Energy loss in the target (per incident particle) | Remarks |
|---|---|---|---|
| Thermal neutrons $^{56}Fe(n,\gamma)^{57}Fe$ | $3\times10^{-4}$ | - | total yields for all levels |
| Deutrons $^{56}Fe(d,p)^{57}Fe$ | $10^{-5}$ | 300 keV | |
| Alpha particles E = 10 MeV, Coulomb excitation | $4\times10^{-8}$ | 3 MeV | the yield is given for excitation of the 136 keV (5/2) level of $^{57}Fe$ |
| $^{57}Fe$ ions accelerated to 100 MeV, then stopped | $10^{-5}$ | 100 MeV | |

Table 4.2 gives an estimate of the excitation probability of $^{57}Fe$ nuclei for various types of bombardment, targets being made of a natural mixture of iron isotopes and having a thickness of about 10 mg/cm$^2$.

As can be inferred from Table 4.2, the excitation of Mössbauer levels in the braked heavy ions is quite competitive with other methods of excitation. A drawback of this approach is the relatively high release of heat in the target per incident particle, which decreases the probability of occurrence of the Mössbauer effect. Also unsolved is the problem of relasing the newly excited particles from the target within a time interval $\Delta t < \tau_0$, so that the working acicular crystal can be grown.

Another work [4.87] proposes the bombarding Mössbauer nuclei could be excited by collisions as they fly through the target without completely stopping in contrast to the above procedure and then precipitated on an external substrate. In flying through the target some of the nuclei will not be excited. These will have to be separated out, which can be done by "magnetic" separation of isomers [4.46] for example.

Estimates [4.87] using $^{57}Fe$ nuclei and a gain factor of $K = 10$ show that about $4.8\times10^{15}$ ions must be in the acceleration beam pulse in order to ensure efficient gamma lasing. In-flight excitation of Mössbauer ions may be most efficient, at least for producing powerful sources of spontaneous gamma emission.

A great deal of interest concerning the passage of particles through crystals was aroused recently by the appearance of KUMAKHOV's paper [4.88] in which it was suggested that powerful spontaneous emission of gamma quanta may be attained from channeled relativistic particles. In a sense, the basic idea of this work is opposite to that proposed by BELYAKOV [4.80], who suggested emission would result from an excited atom rather than the particle itself. According to KUMAKHOV [4.89], the path of the particle would undergo periodic curvature under the effect of the periodic potential, and this is what would cause emission. The intensity (or power) of this emission by a relativistic particle moving through a channel is

$$I = x_m^2 \omega^4 e^2 \gamma^4 / 3c^2 \quad , \tag{4.17}$$

where $x_m$ is the initial vibration amplitude of the particles in the channel, $\omega$ is the emission frequency, e is charge, $\gamma = (1-v_z^2/c^2)^{-\frac{1}{2}}$, and $v_z = dz/dt$ is the particle velocity in the channel (in the case of channeling, the transverse velocity $v_x \ll v_z$).

Since the potential gradients of atomic chains and planes are $10^{11}$-$10^{12}$ eV/cm, the radius of the periodic curvature of the path of the channeled particles is very small, and therefore the intensity I is rather high, greater by a factor of $10^6$-$10^9$ than in modern synchrotrons.

The initiation of this emission has been examined from the standpoint of quantum electronics on the basis of the solution of the Dirac equation for a particle in the harmonic field of atomic planes [4.90]. The emission has been found to be due to spontaneous transitions between levels which occur in the transverse potential of the channel, and it exhibits a number of properties that distinguish it from other known types of emission. It approaches most closely undulatory emission, but differs from it by the dependence of the maximum frequency of the emerging radiation, on energy and mass. For example, the relation $\hbar\omega$ may attain several tens of MeV, which is totally out of the reach of modern synchrotrons. At charged particle energies of 0.1-10 GeV, the emission is most intensive in the range of 0.1 to several tens of MeV. The integrated emission intensity is lower than that for bremsstrahlung, but owing to its greater monochromaticity, the spectral density in the neighborhood of the maximum at $\omega_m$ is greater than in the case of bremsstrahlung.

KUMAKHOV [4.90] suggests the use of the emission from channelized particles to produce an inverted population in a three-level gamma laser [4.55] having an upper level with a short lifetime ($\tau_3 \sim 10^{-10}$-$10^{-13}$ s) and a second level with a sufficiently long lifetime ($\sim$1 s).

KUMAKHOV's work [4.88-90] has been followed by a number of theoretical studies [4.91,92] with results corroborating his idea. It is, however, too early to express great optimism about the possibility of using the emission of channeled particles for pumping Mössbauer transitions because the main problem of line narrowing of levels with long lifetimes has not yet been solved.

In the literature attempts have been made to identify the emission from channeled particles with some other type of emission [4.92], particularly with the coherent bremsstrahlung [4.93] described by SHIFF [4.94] long before the channeling was discovered. KUMAKHOV [4.95] made a detailed analysis of the typical differences between the emission by channeled particles and other types of emission. In this connection we would like to point out BELOSHITSKY's paper [4.96] dealing with emission of electromagnetic radiation by channeled electrons and protons and also containing an analysis of the distinguishing features of this emission.

We do not discuss here some aspects of elastic scattering, diffraction, coherent effects, interference, and bremsstrahlung which occur when fast charged particles move in crystals. For information on these subjects see [4.97-99].

Recent observation of intense gamma emission by electrons having energy $E_0 = 900$ MeV and channelized in a diamond crystal [4.100] gives hope that further new developments will be seen in the near future.

Generally speaking, there are other interactions of charged particles with crystals which should give rise to coherent sources of electromagnetic radiation [4.101, 102] and are less intense than channelized relativistic charged particles. However, no additional analysis of these interactions is required to suggest that further experiments should be conducted on charged particles passing through crystals and the resulting emission analyzed with respect to both scattering angle and frequency.

## 4.8 Non-Mössbauer Gamma-Lasers

The limitations imposed on the gamma-transition energy by the Mössbauer effect have prompted a search for specific physical conditions which would permit the stringent requirements of a Mössbauer experiment to be avoided.

First of all, let us consider the work of WOOD and CHAPLIN [4.29], who proposed making a non-Mössbauer gamma laser by pumping with fast neutrons from laser-fired fusion reaction. As prerequisites for the operation of such a gamma laser, WOOD and CHAPLIN suggested an excited-level lifetime of $10^{-12}$ s, a density of the thermonuclear mixture in the compressed state of $\rho \approx 10^4$ g/cm$^3$, a relative working nuclei ($A \approx 100$) concentration in the thermonuclear mixture of about $10^{-2}$, and an ion temperature of the medium of about 1 keV. To determine the required density of active nuclei in the upper level the authors assumed supercritical conditions and used a formula which is independent of the mean free path ($E_0$) for absorption of gamma quanta by parasitic processes.

However, it is easy to prove by taking these processes into account that the above parameters are inconsistent with attainment of critical (to say nothing of supercritical) conditions.

By using (4.4) for the critical density of working nuclei in the upper level we arrive at the following expression for the non-Mössbauer transition

$$n^* = \left(\frac{E_0}{\pi\hbar c}\right)^2 \frac{\Gamma}{\Gamma_0} \frac{1+\alpha}{\ell(E_0)} \quad . \tag{4.18}$$

In this case, the width $\Gamma$ is a Doppler width proportional to $E_0$. Assuming that the only parasitic interaction involved is Compton scattering of gamma quanta (with a Thomson cross section), we easily deduce from (4.18), using the above parameters, that

$$n^*/n_W \approx 6 \cdot 10^6 \; E_0^3 \tau_{0\gamma} \; . \tag{4.19}$$

Here $n_W$ is the total density of working nuclei, $E_0$ is in keV, $\tau_{0\gamma}$ is the lifetime of the emitting level when in the gamma laser, and $\tau_0$ is the lifetime for spontaneous gamma emission by an isolated nucleus expressed in seconds.

The obvious condition $n^* \leqq n$ immediately suggests that it is impossible to use transitions with $E_0 \gtrsim 50$ keV for gamma lasers with pumping times $t \sim 10^{-12}$ s.

Moreover, it is clearly seen from (4.18) that the limitation on the duration of the transitions used in the gamma laser corresponds to time intervals twice as short as those in the well-known Weisskopf-Moshkovsky formulas, even for the fastest single-particle transitions. In fact, the actual values of $\tau_0$ at $E_0 \approx 100$ keV exceed the calculated values for single-particle electric dipole transitions by at least one order of magnitude [4.103].

Thus, even if only Compton scattering is taken into account, it may be concluded that a non-Mössbauer gamma laser of the type considered in [4.29] is not feasible. It should also be borne in mind that under conditions of high density of the substance and sufficiently low energy of gamma quanta, a more substantial contribution to the parasitic interactions is made by so-called inverse bremsstrahlung.

For the case of complete ionization of the working nuclei, the mean free path of gamma quanta for inverse bremsstrahlung is derived in accordance with [4.104] from

$$\frac{1}{\ell(E_0)} = \frac{4}{3}(2\pi)^3 \sqrt{\frac{2\pi}{3mkT}} \; \frac{z^2 e^6 n_e n_W \hbar^2}{mcE_0^2} \; \frac{5.2\times10^9}{E_0^2} \; \mathrm{cm}^{-1} \quad , \tag{4.20}$$

where e and m are the electronic charge and mass, respectively; $n_e$ is the electron concentration equal to $(z+100)n_W$; $z \approx 50$ is the charge of the working nuclei; and $E_0$ is expressed in keV. For the parameters specified in [4.29], inverse bremsstrahlung already prevails over Compton scattering for $E_0$ as low as 140 keV, while in the range of $E_0 \sim 10$-50 keV this process reduces the value of $\ell(E_0)$ by another factor of 1000-10 below the Compton scattering value. Here the obvious requirement $n^* \lesssim n_W$ is not met. In the opposite extreme case, where no ionization takes place, the situation is similar because of the ordinary photoeffect on the working atoms.

Thus we do not share the optimistic conclusions of [4.29] regarding the possibility of creating a non-Mössbauer gamma laser with a Doppler resonance width and very fast pumping by inelastic scattering of thermonuclear neutrons in a strongly heated medium of extremely high density, or at least not under the conditions stipulated by WOOD and CHAPLIN.

In addition to traditional ways of achieving stimulated gamma emission at nuclear transitions, some investigators [4.105-108] have considered another possibility for providing coherent gamma emission, namely induced annihilation of electron-positron pairs. An electron-positron annihilation emits either two quanta (annihilation of orthopositronium, lifetime $\tau \sim 10^{-10}$ s), or three quanta (parapositronium, lifetime

$\tau \sim 10^{-7}$ s). In the case of two-quantum annihilation, the two quanta are emitted in opposite directions, each one having an energy of $E_\gamma \sim 511$ keV. Just like any electrodynamic process, this annihilation may be induced by the influence of resonance gamma quanta already present in the system. The cross section for induced annihilation is $\sigma \sim \pi ƛ^2$, where $ƛ$ is the wavelength of the emitted gamma quantum and equals $3 \times 10^{-11}$ cm for $E_\gamma \sim 511$ keV. If we take a wavelength at which amplification is e-fold equal to $L \sim 1$ cm, the concentration necessary for the development of a photon avalanche must be $10^{20}$-$10^{21}$ cm$^{-3}$. For a best-case calculation, assuming the use of advanced high-current accelerators or relativistic accumulators, the total number of positrons per pulse is $10^{20}$-$10^{21}$ [4.106]. Therefore, in order to attain the critical concentration, the volume of the working region must be $10^{-8}$-$10^{-10}$ cm$^3$. NAMIOT [4.106] proposed focusing positrons into a medium having a long narrow channel serving as a trap where the newly-formed positronium may build up to the critical concentration. Since the mass of positronium is only $2m_e$, at a concentration of about $10^{21}$ cm$^{-3}$ the positronium gas may become degenerate even at relatively elevated temperatures. The luminescence of the resulting Bose condensate must, according to NAMIOT [4.106], have the natural line width with no Doppler broadening.

The existence of cosmic lasers operating in the SHF (superhigh frequency) and EHF (extremely high frequency) [4.109] naturally calls for analysis of possible shorter-wave stimulated emission processes which might occur under natural astrophysical conditions. That the optical laser effect may manifest itself in the atmospheres of hot stars was pointed out in an earlier work [4.110]. The mechanism of periodic cosmic bursts of gamma rays has yet to be elucidated [4.111], but the existence of induced gamma radiation under astrophysical conditions has been hypothesized [4.21, 22,112].

## 4.9 Main Trends in the Near Future

As shown in this paper, the problem of creating gamma lasers is a complex one and calls for joint efforts of scientists specializing in various fields. Therefore, it is important to outline the research program so as to have a clear understanding not only of the general trend, but also of individual tasks.

It should be pointed out that in the development of gamma lasers we can hardly expect sensational discoveries; first meticulous measurements of various constants characterizing gamma transitions must be made, then experimental methods of isotope separation and crystal growing will have to be developed, and so on. Without these preliminary steps no success can be achieved. In any event, regardless of the scope and deadlines set for accomplishing the main tasks in the program of gamma laser development, the schedule proposed below may stimulate advances in a number of scientific and technological areas and yield many results of intrinsic value.

The following list is far from exhaustive, but may suggest new investigations even at the first steps.

*a) In the Field of Nuclear Spectroscopy*

1) Analysis of the available data and the general systematics of nuclear levels with a view to finding gamma transitions having energies of 5-100 keV (and even up to 200 keV), including transitions between excited levels and transitions in the region of low excitation energies. The levels of interest must be populated during radiative capture of neutrons, the initial targets being either stable isotopes or long-lived (ca.$\sim 10^5$-$10^6$ s) radioactive isotopes.
2) Theoretical and experimental studies of the absolute values and energy dependence of the neutron-capture cross sections for populating levels of interest.
3) Theoretical and experimental studies of the values of isomeric ratios for populating levels of interest by radiative capture of slow neutrons of various energies.
4) Development of new optical laser techniques for spectroscopy of isomeric nuclear states.

*b) In the Field of the Mössbauer Effect*

1) Determination of the probability of the Mössbauer effect after a neutron capture which populates several levels, and of the temperature dependence of this probability for various matrices.
2) Experiments on gamma quanta emitted by targets located in a neutron beam and resonantly absorbed by secondary targets beyond the beam.
3) Experiments on selective population of individual hyperfine components of the lower nuclear levels by combining cryogenic temperatures with strong local or external magnetic fields.
4) Experiments on the production of radio-spectroscopic transitions between the hyperfine sublevels of Mössbauer nuclei (gamma-magnetic resonance).
5) Development of methods for narrowing Mössbauer lines (NMR using internal and external magnetic fields, ultrasonic activation, growing of absolutely perfect crystals, use of coherent effects).

*c) In the Field of Rapid Separation and Collection of Excited Mössbauer Nuclei*

1) Development of hot-atom analysis methods aimed at rapid separation of $n\gamma$-capture products escaping as a result of the Scillard-Chalmers process from the surface of highly dispersed materials into a confined space and their effective chemical binding (e.g., in the form of volatile compounds).
2) Development of effective methods for the rapid separation of chemically-bound excited Mössbauer nuclei in the form of neutral molecules (e.g., by entrainment in a supersonic gas flow) or ions (e.g., by collection in an electric field) and their incorporation into an active emitting system in the form of a film, filament, etc.

3) Development of methods for rapid separation of nuclear isomers by optical laser beams, for example, by selective excitation and ionization.
4) Development of procedures for separating nuclei in the ground and excited states from each other after radiative neutron capture using methods based on the Scillard-Chalmers process and laser excitation of the molecules.

*d) In the Field of Development of Neutron Sources*

1) Development of pulsed sources to produce neutron fluxed of up to $10^{14}$-$10^{15}$ $cm^{-2}$ (if possible, up to $10^{16}$ $cm^{-2}$) within short time intervals ranging from $10^{-6}$-$10^{-4}$ s, for example, thermonuclear sources of the "plasma focus" type.

*e) In the Field of Crystal Physics*

1) Development of techniques for growing crystals with a small number of dislocations.
2) Development of techniques for growing perfect crystals containing atoms with excited nuclei.
3) Development of techniques for rapid growth of microcrystals for the gaseous phase.

*f) In the Field of Laser Problems*

1) Calculations and experiments concerning the kinetics of induced radiation emission in gamma lasers.
2) Calculations and experiments concerning the loss of coherence in gaser beams.
3) Calculations and experiments concerning resonators for gamma (and X-ray) lasers.

*g) In the Field of Astrophysical and Cosmological Problems*

1) Theoretical analysis of possible manifestations of stimulated gamma-radiation processes occurring under natural astrophysical conditions, particularly processes caused by nuclear gamma transitions, positron annihilation, and decay of $\pi^0$-mesons; with due consideration for the various types of non-Mössbauer resonance transitions in which the lines are broadened by recoil and Doppler effects, but very rapid population of the excited level is possible.

*h) In the Field of Laser Spectroscopy of Atoms and Molecules*

1) Theoretical and experimental studies of isomeric shifts and splittings at optical transitions in atoms and molecules.
2) Theoretical and experimental studies of the use of optical-laser radiation to selectively photoionize atoms and photodissociate molecules having excited nuclei.

## 4.10 Possible Applications of Gamma-Lasers

At present it is difficult to define clearly the areas of application of gamma lasers, but their use can be predicted from applications now being made of incoherent X-ray sources and optical lasers.

Let us briefly mention some possible applications. First of all, coherent sources of gamma radiation in the range of wavelengths shorter than 1 Å will be of great importance in nuclear physics as they will permit the study of photonuclear reactions initiated by monochromatic gamma quanta, including reactions with nuclei in short-lived excited states. Next in line are studies of multiple-photon interactions with nuclei and of problems in nonlinear and nuclear optics.

Another application in physics is initiation of multiple-photon interactions with the inner shells of atoms, including induction of transitions of several (deep) electrons in several (different) atoms of a given molecule.

Methods of X-ray optics are already used in studies of the shapes and sizes of micropores, fissures, various defects, and even stress fields caused by dislocations. Coherent gamma radiation may provide for higher resolution in microflaw detection and high-contrast radiography. Another example of industrial use of gamma lasers is microelectronics. The practically zero divergence of the laser beam and its short wavelength will permit producing artwork for microradioelectronic devices measuring less than tenths of a micron.

Another promising area is gamma interferometry and its use in precision measurements of fundamental constants. A combination of optical and interferometric methods may give impetus to the development of holographic techniques for the visualisation of biological macromolecules (DNA, RNA, genes). Although Mössbauer spectroscopy already permits study of the motion of active centers in proteins, the application of gamma lasers to the study of molecular kinetics will trigger a revolution in the natural sciences.

Recent works concerned with the development of gamma lasers and delving into the processes of generation of coherent gamma radiation will stimulate a wide range of studies which are valuable per se.

Examples of results which would have intrinsic value include narrowing of Mössbauer lines, extension of the number of Mössbauer isotopes used in scientific research, and applications of the Mössbauer effect (e.g., to the problems of unit length, gravitational waves, etc.).

## References

4.1 T. Kwan, J.M. Dawson, A.T. Lin: Phys. Fluids *4*, 581 (1976)
4.2 F.A. Hopf, P. Meystze, M.O. Senlly, W.H. Louisell: Phys. Rev. Lett. *18*, 1215 (1975)
4.3 F.R. Arutyunyan, V.A. Tumanyan: Sov. Phys.-Usp. *7*, 339 (1964)
4.4 F.R. Arutyunyan, I.I. Goldman, V.A. Tumanyan: Sov. Phys.-JETP *18*, 218 (1964)
4.5 R.H. Pantell, G. Saicini, H.E. Puthoff: IEEE J. Quantum Electron. QE-*11*, 905 (1968)
4.6 A.V. Vinogradov, I.I. Sobelman: Sov. Phys.-JETP *36*, 1115 (1973)
4.7 L.A. Rivlin: Sov. Biull. Izobretenij *23*, June 25 (1979), application of January 10, 1961
L.A. Rivlin: Vopr. Radioelektron. Ser.1, *6*, 42 (1963)
4.8 B.V. Chirikov: Sov. Phys.-JETP *17*, 1355 (1963)
4.9 W. Vali, V. Vali: Proc. IEEE *182*, 1248 (1963)
4.10 G.C. Baldwin, J.P. Neissel, L. Tonks: Proc. IEEE *182*, 1247 (1963)
4.11 D.F. Zaretskii, V.V. Lomonosov: Sov. Phys.-JETP *21*, 243 (1965)
4.12 J.H. Terhune, G.C. Baldwin: Phys. Rev. Lett. *14*, 589 (1965)
4.13 A.M. Afanas'ev, Yu. Kagan: Sov. Phys.-JETP Lett. *2*, 81 (1965)
4.14 A. Kamenov, Ts. Bonchev: Dokl. Bolg. Akad. Nauk *18*, 12 (1965)
4.15 R.L. Mössbauer: Naturwissenschaften *45*, 538 (1958)
4.16 R.L. Mössbauer: Z. Phys. *151*, 124 (1958)
4.17 R.V. Khokhlov: Pisma Zh. Eksp. Teor. Fiz. *15*, 580 (1972) [Eng. transl.: Sov. Phys. JETP Lett. *15*, 414 (1972)]
4.18 Yu.A. Il'inskii, R.V. Khokhlov: Usp. Fiz. Nauk *110*, 449 (1973)
4.19 V.L. Goldanskii, Yu.M. Kagan: Sov. Phys.-JETP *37*, 49 (1973)
4.20 V.S. Letokhov: Sov. Phys.-JETP *37*, 787 (1973)
4.21 V.A. Bushuev, R.N. Kuz'min: Sov. Phys. Usp. *17*, 942 (1975)
4.22 V.A. Bushuev, R.N. Kuz'min: Appar. Metody Rentgen. Anal. *19*, 71 (1978)
4.23 G. Chapline, L. Wood: Sov. J. Quantum Electron. *6*, 452 (1976)
4.24 G.C. Baldwin: Laser Focus *42* (1974)
4.25 G.C. Baldwin, R.V. Khokhlov: Phys. Today *28*, 32 (1975)
4.26 Yu.A. Il'inskii, R.V. Khokhlov: Izv. Vyss. Ucheb. Zaved. Radiofiz. *19*, 792 (1976)
4.27 G.C. Baldwin: In *Laser Interaction and Related Plasma Phenomena*, Vol.*4A*, ed. by H.J. Schwarz and H. Hora (Plenum Press, New York 1977) p.249
4.28 L.I. Gudzenko, L.A. Shelepin, S.I. Yakovlenko: Sov. Phys. Usp. *17*, 848 (1975)
4.29 L. Wood, G. Chapline: Nature *252*, 447 (1974)
4.30 V.I. Goldanskii, Yu.M. Kagan: Sov. J. Quantum Electron. *6*, 455 (1976)
4.31 V.I. Goldanskii, Yu.M. Kagan: Usp. Fiz. Nauk *110*, 445 (1973)
4.32 V.I. Goldanskii, Yu.M. Kagan: In *5th Int. Conf. on Mössbauer Spectroscopy, Proc.*, Part 3, ed. by M. Hucl, T. Zemčik (Czechoslovak Atomic Energy Commission, Nuclear Information Center, Praha 1975) p.584
4.33 Yu.M. Kagan: In *6th Int. Conf. on Mössbauer Spectroscopy, Proc.*, Part 2, ed. by A.Z. Hrynkiewicz, J.A. Sawicki (Cracow, Poland 1975) p.17
4.34 Yu.M. Kagan: Sov. Phys.-JETP *20*, 243 (1965)
4.35 J.S. Waugh: *New NMR Methods in Solid State Physics* (Mir Publishers, Moscow 1978)
4.36 A.V. Andreev, Yu.A. Il'inskii, R.V. Khokhlov: Sov. Phys.-JETP *40*, 819 (1975)
4.37 Yu.A. Il'inskii, R.V. Khokhlov: Sov. Phys.-JETP *38*, 809 (1974)
4.38 V.I. Goldanskii, S.V. Karyagin, V.A. Namiot: Sov. Phys. JETP Lett. *19*, 324 (1974)
4.39 Yu.M. Kagan: Pisma Zh. Eksp. Teor. Fiz. *19*, 722 (1974) [Engl. transl.: Sov. Phys. JETP Lett. *19*, 373 (1974)]
4.40 A.S. Ivanov, A.V. Kolpakov, R.N. Kuz'min: Sov. Phys.-Solid State *16*, 794 (1974)
4.41 V.A. Namiot: Pisma Zh. Tekh. Fiz. *1*, 113 (1975)
4.42 V.I. Goldanskii, V.A. Namiot: Hyperfine Interact. *3*, 253 (1977)
4.43 G.V.H. Wilson: Appl. Phys. Lett. *30*, 213 (1977)
4.44 C.D. Bowman: "Proposal for Nuclear X-Ray Laser Driven by a Nuclear Explosion", private communication (1975)

4.45 G.A. Askaryan, V.A. Namiot, M.S. Rabinovich: Pisma Zh. Eksp. Teor. Fis. *17*, 597 (1973) [Eng. transl.: Sov. Phys. JETP Lett. *17*, 424 (1973)]
4.46 V.I. Goldanskii, Yu.M. Kagan, V.A. Namiot: Sov. Phys.-JETP *18*, 34 (1973)
4.47 A.V. Podoplelov, T.V. Leshina, Ren.Z. Sagdeev, Yu.N. Molin, V.I. Gol'danskii: Pisma Zh. Eksp. Teor. Fiz. *29*, 419 (1979) [Eng. transl.: Sov. Phys. JETP Lett. *29*, 380 (1979)]
4.48 V.A. Namiot: Pisma Zh. Tekh. Fiz. *1*, 5 (1975)
4.49 J. Borrmann: Phys. Z. *42*, 157 (1941); Phys. Z. *127*, 297 (1950)
4.50 Yu. Kagan, A.M. Afanas'ev: In *Mössbauer Spectroscopy and Its Applications* (IAEA, Vienna 1972) p.143
4.51 Yu.M. Kagan: Sov. Phys. JETP Lett. *20*, 11 (1974)
4.52 A.V. Andreev, Yu.A. Il'inskii: Sov. Phys.-JETP *41*, 403 (1975)
4.53 A.V. Andreev, Yu.A. Il'inskii: Pisma Zh. Eksp. Teor. Fiz. *22*, 462 (1975)
4.54 A.V. Andreev, Yu.A. Il'inskii: Sov. Phys.-JETP *43*, 893 (1976)
4.55 V.F. Dmitriev, E.V. Shuryak: Sov. Phys.-JETP *40*, 244 (1975)
4.56 V.I. Vorontsov, V.I. Vysotskii: Kvant. Elektron. *8*, 69 (1974)
4.57 V.I. Vorontsov, V.I. Vysotskii: Sov. Phys.-JETP *39*, 748 (1974)
4.58 A.V. Andreev: Sov. Phys.-JETP *45*, 734 (1977)
4.59 V.A. Bushuev, R.N. Kuz'min, O.Yu. Tikhomirov: Obrabotka i Interpretatsiya Fizicheskich Eksperimentov *5*, 91 (1976)
4.60 A.V. Andreev, V.A. Galkin, O.Yu. Tikhomirov: Obrabotka i Interpretatsiya Fizicheskich Eksperimentov *6*, 3 (1977)
4.61 V.A. Andreev, Yu.A. Il'inskii, R.V. Khokhlov: Sov. Phys.-JETP *46*, 682 (1977)
4.62 A.V. Andreev, Yu.A. Il'inskii, R.V. Khokhlov: Pisma Zh. Eksp. Teor. Fiz. *3*, 779 (1977)
4.63 G.C. Baldwin, L.E. McNeil: Informal Rpt. LA-7004-MS, US-34, Los Alamos Scientific Laboratory, University of California (1977)
4.64 G.C. Baldwin, B.R. Suydan: Prepr. LA-4R-77-140, Los Alamos Scientific Laboratory, University of California (1977)
4.65 G.C. Baldwin: *Laser Interaction and Related Plasma Phenomena*, Vol.4A, ed. by H.J. Schwarz, H. Hora (Plenum Press, New York 1977) p.259
4.66 V.I. Vysotskii, V.I. Vorontsov: Sov. Phys.-JETP *46*, 27 (1977)
4.67 V.I. Vysotskii: Sov. Phys.-JETP *50*, 250 (1979)
4.68 L.A. Rivlin: Vopr. Radioelektron. *6*, 60 (1962)
4.69 A.V. Kolpakov, R.N. Kuz'min, V.M. Ryabov: J. Appl. Phys. *41*, 3549 (1970)
4.70 R.A. Fisher: Prepr. UCRL-75460, Lawrence Livermore Laboratory (1974) p.2201
4.71 C. Elachi, G. Evans, F. Grunthaner: Appl. Opt. *14*, 14 (1975)
4.72 R.D. Deslattes: Appl. Phys. Lett. *12*, 133 (1968)
4.73 W.L. Bond, M.A. Duguay, P.M. Rentzepis: Appl. Phys. Lett. *10*, 216 (1967)
4.74 R.M. Cotterill: Appl. Phys. Lett. *12*, 403 (1968)
4.75 A.G. Rostomyan, P.A. Bezirganyan: Sov. Phys. Dokl. *23*, 48 (1978)
4.76 V.A. Belaykov, R.N. Kuz'min: In *Mössbauerografia*, Physics Series, Vol.1 (Znanie, Moscow 1979) p.40
4.77 L.A. Rivlin: Sov. Phys. JETP Lett. *1*, 79 (1965)
4.78 L.A. Rivlin: Pisma Zh. Eksp. Teor. Fiz. *13*, 362 (1971)
4.79 L.A. Rivlin: Sov. J. Quantum Electron. *2*, 274 (1972)
4.80 V.A. Belyakov: Pisma Zh. Eksp. Teor. Fiz. *13*, 254 (1971)
4.81 Yu.M. Kagan, F.N. Chukhovskii: Pisma Zh. Eksp. Teor. Fiz. *5*, 166 (1967)
4.82 E.A. Perelshtein, M.I. Podgoretskii: Yad. Fiz. *12*, 1149 (1970)
4.83 A.V. Kolpakov: Yad. Fiz. *16*, 1003 (1972)
4.84 V.A. Belyakov, V.P. Orlov: Phys. Lett. *A41*, 463 (1973)
4.85 E.A. Manykin: Sov. Phys. JETP Lett. *22*, 271 (1975)
4.86 V.I. Goldanskii, F.I. Dalidchik, G.K. Ivanov: Sov. Phys.-JETP *7*, 138 (1968)
4.87 V.A. Bushuev, A.V. Kolpakov, R.N. Kuz'min, Ye.M. Saprykin, D.A. Shelabaev: Vestn. Mosk. Univ. Fiz. Astronomiya *19*, 101 (1978)
4.88 M.A. Kumakhov: Phys. Lett. *57A*, 17 (1976)
4.89 M.A. Kumakhov: Sov. Phys.-Dokl. *21*, 581 (1976)
4.90 M.A. Kumakhov: Sov. Phys.-JETP *45*, 781 (1977)
4.91 V.A. Bazylev, N.K. Zhevago: Sov. Phys.-JETP *46*, 891 (1977)
4.92 N.K. Zhevago: Sov. Phys.-JETP *48*, 701 (1978)
4.93 R.W. Terhune, R.H. Pantell: Appl. Phys. Lett. *30*, 265 (1977)
4.94 L.I. Shiff: Phys. Rev. *117*, 1394 (1960)

4.95 M.A. Kumakhov: Phys. Status Solidi B *84*, 41 (1977)
4.96 V.V. Beloshitskii: Phys. Lett. *64A*, 95 (1977)
4.97 A.I. Akhiezer, Ye.F. Boldyshev, N.F. Shulga: Fiz. Elem. Chastis At. Yadra *10*, 51 (1979)
4.98 A.I. Akhiezer, I.A. Akhiezer, N.F. Shulga: Sov. Phys.-JETP *49*, 631 (1979)
4.99 V.G. Baryshevskii, A.O. Grubich, I.Ya. Dubovskaya: Phys. Status Solidi B *88*, 351 (1978)
4.100 S.A. Vorobyev, V.N. Zabaev, B.N. Kalinin, V.V. Kaplin, A.P. Potylitsyn: Sov. Phys. JETP Lett. *29*, 376 (1979)
4.101 S.A. Akhmanov, B.A. Grishanin: Sov. Phys. JETP Lett. *23*, 515 (1976)
4.102 D. Dialetis: Phys. Rev. *A17*, 1113 (1978)
4.103 E.Ye. Berlovich, S.S. Vasilenko, Yu.N. Novikov: *Vremena Zhisni Vozbuzhdennych Sostoyanii Atomnych Yader* (Nauka, Leningrad 1972) p.232
4.104 Ya.B. Zeldovich, Yu.P. Raizer: *The Physics of Shock Waves and High-Temperature Hydrodynamic Phenomena* (Academic Press, New York 1966)
4.105 V.A. Belokogne: Aviatsiya i Kosmonavtika *6*, 22 (1970); Priroda (Moscow) *7*, 41 (1965)
4.106 V.A. Namiot: "O Positronnykj i Mössbauerovskikj Istochnikach Kogerentnogo Gamma-Izlucheniya", Thesis, Moscow State University (1975)
4.107 L.A. Rivlin: Sov. J. Quantum Electron. *4*, 1151 (1975)
4.108 D. Marcuse: Proc. IEEE *51*, 849 (1963)
4.109 V.S. Strelnitskii: Sov. Phys.-Usp. *17*, 507 (1975)
4.110 N.N. Lavrinovich, V.S. Letokhov: Sov. Phys.-JETP *40*, 800 (1975)
4.111 O.F. Prilutskii, I.L. Rozenthal, V.V. Usov: Sov. Phys.-Usp. *18*, 548 (1976)
4.112 E.A. Arutyunyan: Izv. Akad. Nauk SSSR Ser. Fiz. *39*, 2198 (1975)

# 5. Nuclear Resonance Experiments Using Synchrotron Sources

R. L. Cohen

With 7 Figures

This chapter discusses the problems and possibilities of using synchrotron radiation X-rays to excite low-lying nuclear states. To date "single nucleus" excitations of $^{57}$Fe have been observed. Within the next year, synchrotron-excited nuclear Bragg scattering will probably be observed, opening the path to a number of hyperfine interaction and X-ray diffraction measurements. This chapter deals primarily with the advantages and disadvantages of the ingenious schemes which have been proposed to observe the nuclear Bragg scattering in the presence of the much stronger electronic Bragg scattering.

## 5.1 Overview

Synchrotron radiation (SR) experiments using nuclear resonance can be divided into two categories: those in which the nuclei are acting as individual scattering or fluorescence centers, and those in which interference from scattering due to many cooperating nuclei is dominant. It is convenient to call the former class nuclear fluorescence or absorption experiments, and the latter nuclear Bragg scattering experiments. We will discuss here first the nuclear fluorescence experiments, which are relatively straightforward in their approach and goals, and then discuss the possibilities of the Bragg scattering experiments. At the time of this writing (August 1978), more than a dozen theoretical papers have been written analyzing various aspects of prospective nuclear Bragg scattering experiments, and three experimental groups have mounted programs to implement such experiments. However, the nuclear Bragg scattering of SR has not yet been observed. Nuclear resonance absorption of SR has been experimentally observed in one recent experiment [5.1] using $^{57}$Fe.

It is interesting to note that despite the enthusiasm with which physicists are pursuing these "new" experiments, both the Bragg scattering and fluorescence experiments have been realized for almost twenty years using conventional excitation sources. A conventional X-ray tube source was used [5.2] to excite and observe nuclear resonance fluorescence in $^{19}$F and $^{55}$Mn. that experiment was preceded by others [5.2] using the bremsstrahlung from betatron-accelerated electrons, and gamma rays from nuclear reactions [5.3]. At about the same time, nuclear Bragg scattering was being observed [5.4] using $^{57}$Fe, with radiation coming from a source of $^{57}\mathrm{Co} \rightarrow {}^{57}\mathrm{Fe}$. All of these early experiments were both extremely difficult and very limited in the information they supplied. They did not have great impact on the mainstream of nuclear and solid state physics and did not lead to extensive further work. Why then this sudden enthusiasm for similar experiments with SR sources? The answer, basically, is that the unique characteristics of synchrotron radiation – the collimation, the polarization, the pulsed nature, and the intensity – may make possible experiments which are qualitatively different from those previously carried out. Only time will tell whether the severe experimental difficulties involved in this new series of experiments are justified by the new insights gained.

Many of the problems faced in the implementation of nuclear SR experiments can be seen by examining the formula for the cross section for the absorption of a photon of energy E by a nucleus:

$$\sigma(E) = \frac{1}{2\pi} \frac{h^2c^2}{E_0^2} \frac{2I_e+1}{2I_g+1} \frac{1}{1+\alpha} \left[1 + \frac{4(E-E_0)^2}{\Gamma^2}\right]^{-1} , \tag{5.1}$$

where $E_0$ is the transition energy, $\alpha$ is the internal conversion coefficient of the transition, $\Gamma = 1/\tau$ is the uncertainty principle width of the state, derived from $\tau = 1.44t_{1/2}$, and $I_g$ and $I_e$ are the spins of the nuclear ground and excited states.

Some of these values are given in Table 5.1 for some states which might be used for SR experiments. It is important to realize that the cross section which determines the strength of nuclear resonance scattering from broad-band sources is the energy-integrated cross section, and is thus proportional to $\Gamma$, so that the best states for such experiments would appear to be short-lived. The time resolution of

Table 5.1. Some nuclear transitions which are promising for synchrotron radiation experiments. A half life of 1000 ps corresponds to a resonance of $4.6 \times 10^{-7}$ eV

| Isotope | Transition energy [keV] | Half-life [ps] | Peak resonance cross-section [$10^{-20}$ cm$^2$] | Natural abundance [%] |
|---|---|---|---|---|
| $^{19}$F | 110 | 600 | 20.0 | 100 |
| $^{57}$Fe | 14.4 | 97800 | 256.0 | 2.2 |
| $^{83}$Kr | 9.4 | $0.14\times10^6$ | 107.0 | 11.6 |
| $^{157}$Gd | 64 | $0.46\times10^6$ | 23.0 | 15.7 |
| $^{159}$Tb | 58 | 105 | 10.5 | 100 |
| $^{165}$Ho | 95 | 22 | 8.3 | 100 |
| $^{181}$Ta | 6.2 | $6.8\times 10^6$ | 167.0 | 100 |
| $^{181}$Ta | 136 | 40 | 6.0 | 100 |
| $^{187}$Re | 134 | 10 | 5.4 | 63 |
| $^{201}$Hg | 32.2 | 200 | 0.95 | 13 |

detectors is limited, however, so that for states of half-life shorter than about 20 ns, it is not possible to discriminate between the scattering from electrons and the scattering from nuclei by their different time dependences. As will be seen below, this discrimination is a vital part of all of the experimental arrangements which have been proposed to date.

Significant interference arises from the direct electronic scattering of the SR by the atomic electrons and from X rays following core-level photoelectric absorption. Although cross sections for these processes are much smaller than the peak cross sections for the nuclear resonances, the electronic processes are broad-band rather than being sharply tuned. If the synchrotron radiation is filtered to $\sim 1$ eV bandwidth before carrying out the nuclear scattering experiment, the electronic scattering processes will act on *all* the X rays, while the nuclear scattering will be effective only for a narrow ($\sim 10^{-8}$ eV wide) energy slice of the 1 eV wide radiation. The electronically scattered radiation can easily be many orders of magnitude larger than the nuclear scattering we want to investigate. This interference has been a persistent problem in nuclear resonant scattering experiments. Already in 1962, SEPPI and BOEHM [5.2] pointed out: "This problem might be solved by taking advantage of the instantaneous character of the atomic scattering as compared to the relatively long lifetime of the nuclear excited states. Through the use of a pulsed X-ray beam and a properly gated detector it should be possible to observe only the nuclear excitation events in the sample."

With the inherently short pulses of synchrotron radiation, it is relatively easy to implement this approach by detecting only delayed photons emitted by the relatively long-lived nuclear excitations. To use this discrimination, however, one must study nuclear states with lifetimes of at least 20 ns or so in order to have adequate time separation between the electronic and nuclear scattered radiations. Thus, very short-lived states, which have the largest resonance widths, cannot be studied this way. In fact, all of the currently planned experiments use the 14.4-keV transition in $^{57}$Fe with a 100 ns half-life, although the energy-integrated cross section is much smaller than that of many other candidates.

Another approach taken to increase the effective ratio of nuclear-to-electronic scattering is the use of nuclear Bragg scattering. Experiments of this kind are discussed in Sect.5.3.

## 5.2 Single Nucleus Excitations

### 5.2.1 Mössbauer Effect

The Mössbauer effect, more formally known as "recoil-free gamma-ray resonance absorption", was first reported in 1958. Since then, experimental research with the use of the Mössbauer effect has diffused rapidly into the diverse fields of solid state physics, metallurgy, chemistry and biochemistry [5.5].

In conventional Mössbauer spectroscopy, the experimental arrangement may be described as follows. An atomic nucleus makes a transition from an excited state to its ground state emitting a gamma ray. This gamma ray has approximately the right energy to be resonantly absorbed by a nucleus of the same kind in its ground state. Small perturbations in the energy of nuclear levels in the absorber can be measured by observing the change in gamma-ray energy required for the gamma ray to be resonantly absorbed. The nuclear hyperfine (HF) interaction and isomer shifts measured in this way provide useful information about the environment of the atom under study.

Although a number of claims have been made that SR could replace conventional sources for Mössbauer spectroscopy experiments, it is my feeling that this is extremely unlikely. The convenience and low cost of radioactive sources will make them difficult to displace. Although there are a very few possible Mössbauer levels which cannot easily be reached from radioactive parents, most of those nuclear states are undesirable or uninteresting for some other reason in any case. It is my feeling that Mössbauer experiments using SR will be worthwhile only where the unique characteristics of SR provide basically new possibilities.

RUBY [5.6] has already proposed the most straightforward approach to synchrotron Mössbauer experiments. His proposal is to look at fluorescent de-excitation radiation (X rays, gammas, or conversion electrons) from the Mössbauer absorber in a scattering geometry. By gating the counters *off* during the radiation pulse, ($\sim 10^{-10}$ s) all of the Rayleigh, Compton, and X radiation (following photoeffect) should be eliminated as they are prompt (on the scale of $10^{-10}$ s). Thus, provided that the Mössbauer absorbing state had a lifetime of $\sim 1$ ns or longer, all of the observed radiations should arise from gamma-ray absorption exciting the nuclear state. This allows one to detect the occurrence of gamma-ray resonance absorption, but does not allow Mössbauer spectroscopy because the 1 eV bandwidth of the monochromator is much greater than the nuclear absorption width, $\sim 10^{-8}$ eV. An immediate useful result of this configuration is that one has a simple, absolutely stable monochromatic detector with a linewidth orders of magnitude smaller than that of the crystal monochromator. Since count rates from 1 to $10^4$ c/s could be obtained, the nuclear resonance detector would be an excellent tool for investigating and improving the characteristics of the monochromator.

To perform Mössbauer resonance experiments despite the broad monochromator output spectrum, RUBY [5.6] proposed (Fig.5.1) putting a "notch filter" in the beam, upstream of the fluorescing absorber, as was done in the old gamma-ray resonance fluorescence experiments [5.3]. This filter would be another Mössbauer absorber, whose absorption energy could be varied by Doppler modulation, as in a conventional Mössbauer experiment. When the energy of the line in the filter was the same as that of a resonance absorption line in the absorber, a dip in fluorescence counting rate

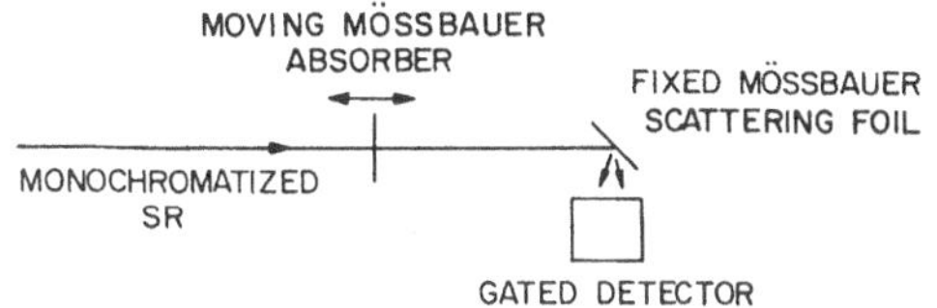


Fig. 5.1. Basic configuration of Mössbauer experiments using SR. X rays in the incoming beam within the resonance width of the nuclear transition are absorbed by the nuclei in the fixed foil, and secondary radiations from the excited nuclei are observed by the gated detector. The detector is turned on after the SR flash to make it insensitive to the (prompt) electronically scattered SR. The moving absorber acts as a sharp adjustable filter to absorb narrow regions of the incoming beam energy [5.6]

Table 5.2. Comparison of radioactive sources and storage rings for nuclear resonance experiments

| Trait | Radioactive source $^{57}$Co, 30 mCi, \$1 K | Storage ring SSRL, measured 1 mrad, 33 mA, 3.7 GeV | Storage ring, with wiggler, BNL, calculated 2.5 GeV, 500 mA |
|---|---|---|---|
| 14-keV photons/s, solid angle | | | |
| 2 eV, $10^{-2}$ sr | $10^5$ | $10^{10}$ | $10^{13}$ |
| $10^{-8}$ eV, $10^{-2}$ sr | $10^5$ | 50 | $5\times10^4$ |
| $10^{-8}$ eV, $10^{-7}$ sr | 1 | 50 | $5\times10^4$ |
| Availability | 100% | 1% | 10% |
| Signal/Background | 1 | $10^{-8}$ | $10^{-8}$ |
| Polarization | Possible | Ideal | Ideal |
| Timing | Very limited | Ideal | Possible |

would occur. By varying the filter energy, the absorption cross section of the absorber could be plotted out as a function of energy.

A quick calculation will demonstrate that for conventional Mössbauer experiments, even neglecting certain restrictions which will be discussed below and assuming that detectors could be perfectly gated and would not be saturated by the electronically scattered synchrotron radiation, this technique will not replace the use of radioactive sources (Table 5.2). At present, under good SSRL operating conditions, up to $10^{10}$ photons/s-eV at 10 keV are available after a channel-cut crystal monochromator which subtends 1 mrad. In the normal $10^{-8}$ eV linewidth of an $^{57}$Fe Mössbauer experiment, we would then have about $10^2$ photons/s, much less than what a conventional source would produce. (A 50 mCi $^{57}$Co source, costing ~\$2000, gives ~$10^5$ 14-keV photons/s into a solid angle of 0.01 sr.) The increased flux that will be available with new monochromators and wigglers may result in a thousand-fold increase in counting rate. This intensity is still only comparable to that from the radioactive source. The situation may be changed if helical wigglers provide two orders of magnitude further increase.

For the present then, the question is whether the other characteristics of synchrotron radiation make it possible to do experiments which cannot be done with conventional sources. There are three characteristics of synchrotron radiation which suggest interesting possibilities: the polarization, the pulsed nature, and the good collimation (Table 5.2).

*Polarization:* Polarized Mössbauer gamma rays have been proposed and, in a few cases, used [5.7] for determining the direction of magnetization and sublattice magnetizations. These experiments can be performed with conventional sources by filtering out some of the hyperfine lines of a split source; the radiation remaining is partly polarized. This is accomplished at a considerable loss in intensity. The availability of "free" polarization would undoubtedly make such experiments relatively more attractive and competitive with neutron scattering.

*Pulsed Beam:* There has occasionally been discussion of Mössbauer experiments to measure the hyperfine structure of ions in metastable, low-lying, optically excited states, typically of lifetime $10^{-3}$ s. These experiments have not been feasible in the past. Even though it is possible to put a large fraction of the ions into the optically excited state with a strong light pulse, they quickly decay to the electronic ground state. The power required to keep a large fraction of the ions in the excited state continuously is too large to supply easily. If, however, the light were pulsed to do the optical excitation in synchronization with the synchrotron radiation pulses, the Mössbauer experiment could take place before the excited ions had decayed. Other experiments could be based on the electronic excitation of the solid by the synchrotron radiation pulse itself. For example, Fe or Sn could be doped into Si, and the effects on the Mössbauer spectrum of electrons excited into conduction states could be observed. This could provide direct measurements of trapping times. Using time-of-flight delay techniques for the Mössbauer synchrotron radiation, it would also be possible to make Mössbauer measurements as a function of time after an initial exciting synchrotron radiation pulse. In insulators and semiconductors, a rich variety of effects would be expected to arise. All these experiments would be impossible with radioactive sources.

*Collimation:* The very good collimation of the synchrotron radiation suggests that certain Mössbauer experiments, which are severely limited by solid angle considerations, would be greatly eased by the use of synchrotron radiation. One such experiment is the gravitational red shift measurement, where the size of the effect is proportional to the distance between the source (at the base of a tower) and the absorber (at the top); some compromise must be reached between the increasing effect and decreasing solid angle as the tower height increases (see Chap.3). For current synchrotron intensities and a 100-m tower, the synchrotron radiation beam is equivalent to that from a source of 0.25 Ci, which is realizable without enormous expense. Anticipated improvements in SR intensity using wigglers would make a synchrotron source practical for these experiments. The equivalence between gravitational and inertial mass, now established for gamma rays to ~10% by conventional Mössbauer experiments [5.8] could probably be improved by two orders of magnitude by such experiments [5.9].

The inherently good collimation of SR would also facilitate measurements of the interference effects between nuclear and electronic Bragg scattering which allow direct determination of the X-ray scattering phase factors [5.10-12]. A few such experiments have been carried out using Mössbauer radioactive sources [5.10,11], but they have been extremely difficult. The main limitation is that because of the small acceptance angle of a Bragg reflection, the count rate into a Bragg peak using a radioactive source is very small, typically 1/s to 1/min. This problem would be ameliorated by the use of SR sources, which have angular divergence comparable to the acceptance angle of the crystal.

*Mössbauer experiments in which radioactive sources cannot be used:* In a few cases, there is no easily available radioactive parent for the level to be studied, or the obvious source produces severe interfering lines. Examples of the former are $^{19}F$, $^{40}K$, $^{61}Ni$, $^{121}Sb$, $^{157}Gd$ and $^{201}Hg$; of the latter, $^{73}Ge$, $^{181}Ta$, and $^{169}Tm$. Thus, Mössbauer studies with a large number of isotopes could be carried out more easily using synchrotron radiation than using conventional sources. Research using radioactive Mössbauer absorbers (i.e., long-lived radioactive species) would also be easier with SR since gated counters and a small solid angle could be used to reduce the background from the decay of the absorber nuclei.

### 5.2.2 Nuclear Excitation Without the Mössbauer Effect

One of the major difficulties to be expected in the "Mössbauer experiments without radioactive sources" is that nuclear resonant excitations will arise not only from zero-phonon processes (Mössbauer effect), but also from one- and multi-phonon excitations, because the exciting radiation bandwidth is much wider than the phonon spectrum. Since these processes will result in nuclear excitation indistinguishable from that of the Mössbauer absorption, only f (the recoil-free fraction, or Debye-Waller factor) of the emitted secondary radiation will arise from zero phonon absorptions. Thus, the resonance intensity is diluted, just as in conventional experiments, by f.

Since the absorption line for non-recoil-free transitions is inhomogeneously broadened by the phonon spectrum no useful hyperfine information can be obtained from that absorption. It is, however, possible to take advantage of this effect and derive hyperfine information from the precession rate of the excited nuclear state [5.13]. This would essentially be a cross between a perturbed angular correlations (PAC) experiment, where the precessing state is fed by a gamma decay, and implantation perturbed angular correlation technique (IMPACT) experiments, where a charged particle impact provides both the excitation and the initial orientation axis of the system. SR excitation experiments would have the advantages of both these older techniques: the recoil impact would be small, so the nucleus under study would not be displaced, and the size of the gamma fluorescence anisotropy would be very large, since the exciting synchrotron radiation is polarized. The precession could be detected as a time-integral or time-differential effect. The 1 μs spacing between SR pulses would be ideally matched to the properties of the nuclei of interest.

Another means of detection that has recently been used with great success at the Hahn-Meitner Institute is the "stroboscopic" technique [5.14]. In that approach, in which the incoming beam is pulsed at a fixed frequency, the applied magnetic field (which provides the hyperfine field) is varied to make the precession frequency of the nuclear excited states equal to the repetition rate of the beam pulses so that all the nuclei are precessing in phase. This produces an extremely high sensitivity to small hyperfine field perturbations, and it is easy to measure, for example, the Knight shift in a metal.

Excited state double resonance can be carried out by impressing an RF field on the excited nuclei and inducing transitions between the nuclear hyperfine levels [5.15]. The change in substate populations would be observable via the altered anisotropy of the emitted radiation. These experiments are also difficult to carry out with radioactive sources [5.15], but could become much more attractive with strong SR sources.

The study of hyperfine interactions via the perturbed angular correlation between the absorbed and emitted photons would be a very powerful tool because it would be freed from the limitations of the recoil-free fraction. Thus, medium-energy transitions (50-100 keV) could be studied at high temperatures, and paramagnetism and nuclear relaxation times could be studied in liquids. The only requirement would be a nuclear state at a low enough energy to be excited by the synchrotron radiation and a long enough lifetime to precess noticeably before the decay.

If the incident SR has beem monochromatized to a 1 eV bandwidth, the ratio of nuclear scattering to electronic scattering for most of the transitions of interest for perturbed angular correlation work is only about $10^{-6}$. Thus, realization of the experiments discussed above will require a radical improvement either in detector design (particularly in time resolution and reduced afterpulsing) or in monochromator design (to provide a narrower energy bandwidth for excitation). Both these areas are discussed below.

### 5.2.3 Experimental Results

In contrast to the many possibilities and hopes for SR in exciting nuclear resonance the experimental results to date have been very limited. The only observation of such an effect reported as of this writing has been the experiment of COHEN et al. [5.1] (Figs.5.2,3). In that work, a gated detector was developed to observe the conversion electrons produced from the 14-keV state of $^{57}$Fe excited by SR. In principle it should be easy to observe this effect, since even currently available SR beams can put ~10 photons/s within the linewidth of the $^{57}$Fe resonance, and the nuclear cross section is large enough to absorb all of these photons. The principal difficulty arises from the scattering by the atomic electrons which interact with the entire width (1-5 eV) of the monochromatized incident SR and thus produce scattering several orders of magnitude stronger than that from the nuclei. An obvious ans-

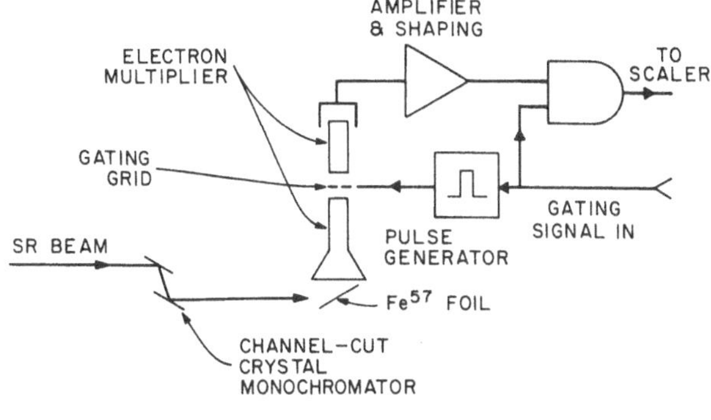


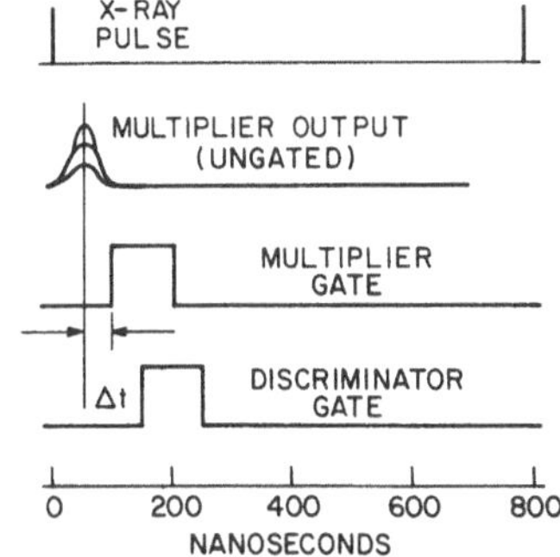


Fig. 5.2. *Top* — Schematic drawing of gated conversion-electron detector used at the Stanford Synchrotron. *Bottom* — Timing diagram for the gating circuits shown above. The timing sequence is repeated at the 1.28 MHz rotational frequency of the synchrotron

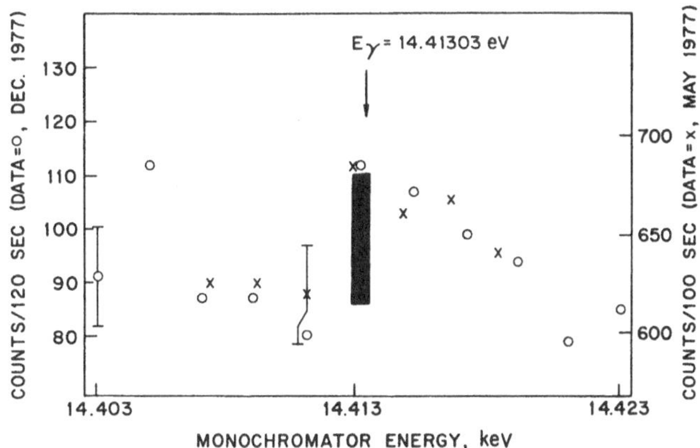


Fig. 5.3. Resonance curves observed by scanning the X-ray monochromator over the nuclear excitation energy. The hyperfine-broadened nuclear resonance is $5 \times 10^{-7}$ eV wide, so the shape of the observed resonance actually represents the monochromator pass function, about 3 eV wide. Actual numbers of counts are shown (for 100 s (X) or 120 s (O) counting times) for two different runs. The vertical scales have been selected to merge the data points. Changes to the detector between the two runs account for the reduced background and slightly lower efficiency in the second run. The vertical bar shows the position and calculated size of the resonance under these operating conditions

wer to this would be to take advantage of the time separation between the (prompt) electronic scattering from the electrons and the delayed reradiation ($t_{1/2}$=100 ns) from the excited nuclei as was discussed above. This procedure, however, has two severe problems associated with it. First, the lifetime of the nuclear state must be at least comparable to the recovery time of the detector (e.g., 10-50 ns for fast scintillators with photomultipliers, or for semiconductor detectors) so that only states with long lifetimes, and therefore small integrated cross sections, are accessible. Secondly, radiation detectors are not perfect – afterpulsing in electron multiplier devices and late collection of trapped charge in semiconductor devices are long standing problems of radiation detectors. In resonant scattering of SR, each radiation pulse would give several photons scattered into the detector by electronic processes, whereas only a few (delayed) nuclear scattered pulses per second would be expected. Thus the electronically scattered pulses are perhaps $10^6$ times as frequent as the delayed nuclear pulses, and if the detector produces a spurious afterpulse for each $10^2$ detected photons, these afterpulses will overwhelm the nuclear resonance scattering signal.

In the light of these problems, the detector design shown in Fig.5.2 incorporates the following features:

1) Conversion electrons arising directly from the nuclear state are detected. This provides increased intensity over the detection of reradiated photons and also avoids the use of scintillation materials which have long "tails" on the fluorescence.
2) The electron multiplier is gated off during the SR pulse so that the prompt electronically-scattered radiation (photoelectrons in this case) is not amplified. Thus, the afterpulsing is virtually eliminated because the primary pulses are not propagated down the multipler. It is impractical to gate the electron multiplier input by retarding potential or grid techniques because the electron energy spectrum extends to 14 keV.
3) The channeltron multiplier used has an inherently low afterpulsing ratio because of its small volume and (in the second stage) helical structure. This property is further enhanced by using a low-noise charge sensitive preamplifier to detect the amplified electron pulse so that only a relatively low electron-multiplier gain was needed.

The detector design shown in Fig.5.2 has been used at SSRL on a standard experimental beam line, and the resonance curve shown in Fig.5.3 was produced. A flux of $2\times10^9$ 14-keV photons/s was incident on the detector foil with pulses occurring at a 1.28 MHz repetition rate. Approximately 20 nuclear excitations per second were produced, leading to a count rate of $\sim$0.2 c/s after accounting for the solid angle, detector efficiency, and gating fraction. About 5 prompt photoelectrons per beam pulse were incident on the electron detector. The multiplier and discriminator gating reduced the prompt feedthrough and afterpulsing from these photoelectrons by a factor of approximately $10^6$, allowing the relatively weak nuclear resonance absorption signal to be observed as the monochromator was tuned through the nuclear resonance energy. The hyperfine-broadened nuclear resonance and photon sidebands are very narrow compared to the monochromator passband, and the observed resonance curve is actually a profile of the monochromator function.

Although the detector discussed here could be used for studying monochromator profiles and establishing absolute energy calibrations, it will probably be more useful as a low-background detector in conjunction with the nuclear Bragg scattering experiments discussed in the next section. Since the nuclear resonance cross section is ~500 times larger than the photoelectric cross section, a substantial increase in discrimination against the electronic Bragg scattered radiation can be obtained.

## 5.3 Nuclear Bragg Scattering

### 5.3.1 Proposed Experiments

When the outgoing scattered wave fronts from individual nuclei in a crystal constructively interfere, the scattering becomes a cooperative phenomenon, and the scattered intensity is greatly enhanced. This effect is exactly analogous to electronic Bragg scattering and thus has been called nuclear Bragg scattering. The analogy is close enough that many of the concepts well developed to understand X-ray scattering in crystals can be used. However, a true analysis of the nuclear scattering case is considerably complicated by the resonant nature of the nuclear scattering cross section [5.11,12,16,17], and analysis of a realistic case must include the nuclear hyperfine structure which divides the nuclear levels into a number of close-lying substates. Many articles have been published recently analyzing the results to be expected in SR-excited nuclear Bragg scattering experiments [5.18-24]. The main aim of this section will be to discuss the proposed experiments, emphasizing the compromises and problems which make the proposals so difficult to realize.

Directly comparing the nuclear and electronic cases, the following general statements can be made. All values cited below are for the 14-keV level of $Fe^{57}$, which has so far been proposed for all nuclear Bragg scattering experiments.

1) The nuclear scattering cross section is about twenty times larger than the electronic cross section so that the penetration depth of the radiation is much smaller in the nuclear case, and the "nuclear Darwin width" is much greater.
2) The energy transmission function of a nuclear Bragg "monochromator" is not determined by the nuclear Darwin width, but by the very narrow inherent width of the nuclear resonance ($\Delta E/E=3\times10^{-13} \sim 10^{-8}$ eV for $Fe^{57}$).

3) Since the nuclear excited states are relatively long-lived, the time dependence of the nuclear Bragg scattered radiation is not simply that of the input radiation, but is much more complex. It is dependent not only on the collective nature of the excited state inside the scattering crystal, but on the nuclear hyperfine interactions [5.18,24].

In standard (radioactive source) nuclear Bragg scattering experiments, the largest single problem is intensity – a radioactive source emits relatively few photons/s into the solid angle accepted by a crystal Bragg reflection. In SR experiments, as pointed out above, adequate flux is readily obtained from the high intensity (SSRL, DORIS, VEPP-4) storage rings. However, in SR nuclear Bragg scattering experiments as in the fluorescence experiments discussed above, the main problem is that monochromatization of the SR by electronic Bragg scattering leaves a beam which is still very wide in energy in comparison with that of the nuclear resonance linewidth. Thus, only a very small fraction (typically $10^{-8}$) of the incoming beam will be subject to nuclear Bragg scattering. A number of ingenious schemes have been suggested to enhance the fraction of nuclear Bragg scattered radiation in the emergent beam. The main aim of these proposals has been to find ways to suppress the electronic Bragg scattering, which inherently occurs at the same angles as the nuclear scattering. It is possible that a number of these techniques may eventually be combined to provide increased discrimination, at the cost of increased complexity. It also appears that timed detectors (as discussed above for the nuclear fluorescence experiments) may be used to allow discrimination between the nuclear and electronic scatterings on the basis of their different time dependences.

Here are brief descriptions of some of these schemes:

1) Single nuclear Bragg scattering with gated (timed) detector to discriminate against electronic Bragg scattering (Fig.5.4). Use of a Bragg angle near $90^\circ$ provides high dispersion and reduces the effective energy width of the electronically scattered radiation [5.20].
2) Two-crystal nuclear Bragg scattering, with misalignment of the electronic Bragg component [5.20]. Since the nuclear Darwin width is so much greater than the electronic, it should be possible to slightly misalign the crystals from the ideal Bragg angle and still transmit the nuclear reflected radiation, but attenuate the electronic scattering (Fig.5.5).
3) Use of a polyatomic crystal in which the electronic Bragg scattering factor is very small for a particular reflection. If only one set of lattice sites in a crystal is occupied by nuclear scattering centers, it may be possible to find reflections for which the electronic scattering factor is close to zero. If the nuclear Bragg reflection does not suffer from the same cancellation, it should thus be possible to get a strong nuclear scattering factor and weak electronic one at the same Bragg angle. The following three approaches have been proposed; of these, the second and third have already been demonstrated using Mössbauer source ($^{57}$Co) excitation.

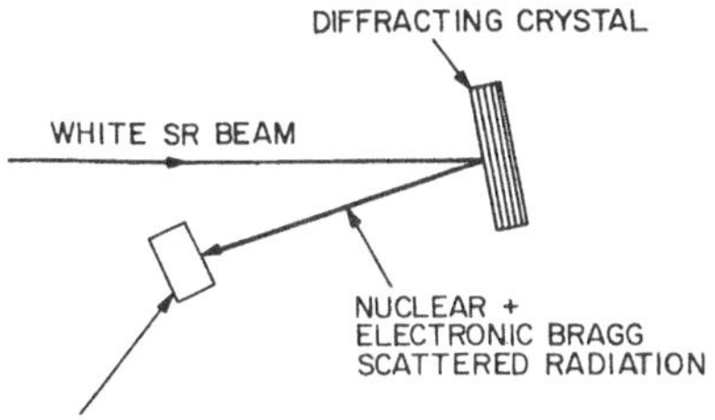


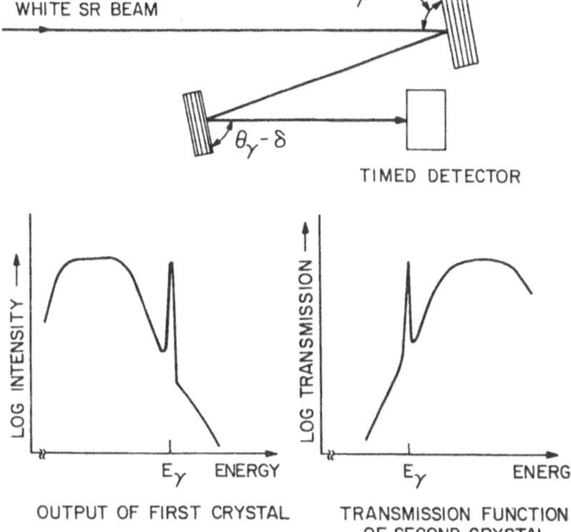


▲ Fig. 5.4. Nuclear Bragg scattering experiment using timed resonant detector to provide both energy and timing discrimination between nuclear and electronic Bragg scattered radiation

Fig. 5.5. Double nuclear Bragg scattering experiment to increase discrimination against electronic Bragg scattering. The crystals are misaligned by $\delta$ from the scattering angle $\theta_\gamma$ corresponding to the nuclear transition energy. If $\delta$ is larger than the electronic Darwin width, but smaller than the nuclear Darwin width, the electronic reflection will be greatly attenuated, but the nuclear reflection will have almost the full intensity [5.20]

For FeTi, a CsCl-structure-ordered intermetallic, odd order reflections have almost zero intensity because of the near-cancellation of the scattering from the Fe and the Ti atoms [5.20]. For $^{57}$Fe nuclei on the Fe superlattice, of course, this cancellation does not occur. Thus, the electronic Bragg scattering is much weaker than the nuclear Bragg scattering. This "enrichment" of the nuclear-scattered part of the beam can be cascaded by additional reflections (Fig.5.5).

In $K_4Fe(CN)_6 \cdot 3H_2O$, BLACK and DUERDOTH [5.10] have shown that for the (080) reflection, the electronic Bragg scattering is negligible due to accidental cancellation of the electronic scattering from the iron ions and all the other ions in the crystal. The nuclear Bragg scattering was readily observed over the very weak electronic Bragg scattering. In some cases it might be possible to improve this accidental cancellation either by varying the crystal temperature to alter the relative Debye-Waller factors of the ions whose scattering is cancelling, or by partial replacement of one ion (e.g., $K^+$) by a chemically similar ion (e.g., $Na^+$) with a different scattering cross section to adjust the net scattering from one sublattice.

The Russian nuclear Bragg scattering research group is planning [5.21] to use the (777) reflection from $\alpha$-$Fe_2O_3$, a reflection which is symmetry-forbidden for the electronic Bragg scattering, but allowed for the nuclear Bragg scattering. This apparent contradiction can be explained as follows. The nuclear levels of $Fe_2O_3$ are split by the hyperfine interaction because the crystal is magnetically ordered, being a canted antiferromagnet at room temperature. The magnetic structure gives rise to four magnetic sublattices for the iron ions with different axes for the hyperfine interaction. Thus, the scattering from the nuclear hyper-

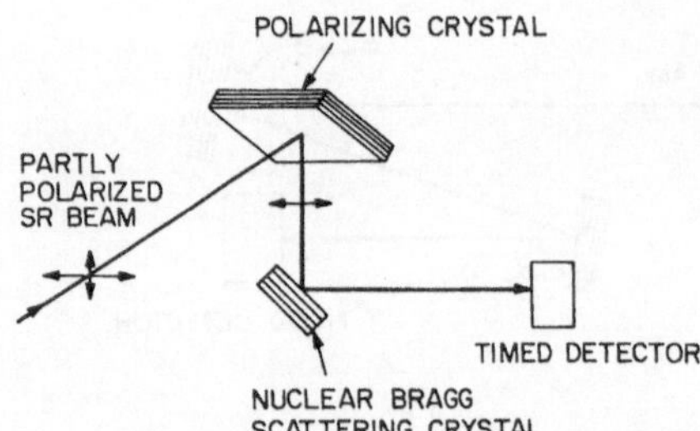


Fig. 5.6. Use of the polarization dependence of the electronic scattering to reduce the electronic scattering. If the second scattering is at a Bragg angle near 45°, the electronic scattering is very weak. The first scattering is primarily to increase the polarization of the SR [5.19]

fine levels sees a lower symmetry than that of the crystal and becomes an allowed reflection [5.25]. Put in a slightly different way, the magnetic structure provides a "hyperfine superlattice" so that the nuclear Bragg reflection would correspond to the superlattice lines of a lower symmetry structure. A suppression of the electronic Bragg scattering by a factor of $>10^5$ has been claimed for this approach [5.21].

4) Use of a thin crystal [5.19]. Since the nuclear cross section is so much larger than the electronic, a crystal containing only a few hundred layers of atoms provides good nuclear Bragg scattering intensity, while the electronic Bragg scattering is still relatively weak.
5) Use of the different multipolarities of the electronic and nuclear scattering [5.19] (Fig.5.6). Since the electronic scattering is E1 while the nuclear scattering is M1, the angular dependences of the differential scattering cross sections are different because of the polarization of the SR. In particular, if the incoming radiation is fully polarized, and the beam is scattered at 90° in the plane of polarization by a Bragg scattering at a 45° Bragg angle, the electronic Bragg scattering is greatly reduced, but the nuclear scattering will still be strong. To make this approach effective, the SR must have its polarization increased (for example, as shown in Fig.5.6) by one or more Bragg scatterings.

It should be emphasized that many of these ideas can be combined to increase the discrimination against the electronic scattering. However, all of the proposals depend strongly on the perfection of the scattering crystals to eliminate the electronic scattering. Thus, an approach which may appear best in principle may be inferior in practice since the particular crystal required may be unavailable. Because of the uncertainty arising from these materials science problems, it is difficult to anticipate which of these options will be the most effective and how many layers of complexity will be required to obtain a useful result.

### 5.3.2 Experimental Problems

The basic problem of SR nuclear Bragg scattering experiments is that there are $\sim 10^8$ times more photons that can be electronically Bragg scattered. The wealth of ingenious ideas outlined above suggests that this basic problem can be overcome and that the success of these experiments is assured. However, there are a number of experimental problems which impede the realization of the approaches mentioned above. Most of these problems arise from the fact that the $^{57}$Fe resonance is the one used. Unfortunately (see Table 5.1), the only other isotope having a long enough half-life to do time-resolved measurements and still having a reasonable energy-integrated cross section is $^{83}$Kr. The obvious materials science problems of making and using diffracting crystals of this isotope seem to have discouraged serious consideration; it remains to be seen whether the obvious problems with krypton are more difficult to overcome than the less obvious problems with iron.

Using $^{57}$Fe, the most obvious problem is that virtually all high-iron materials are magnetic at room temperatures so that the nuclear resonance is split by the hyperfine interaction into (normally) 6 lines. Thus the effective resonance scattering cross section is reduced by a significant factor ($\sim 4\times$), and the Bragg-scattered intensity is reduced by $4^2$, or by 16. Another obvious problem is the low isotopic abundance of $^{57}$Fe, only 2.2%. Thus, not only must special crystals be grown for these experiments, but they must be grown from exceedingly scarce and expensive ($10/mg) raw material.

Although use of pure iron crystals would appear to be the most direct path, there are additional barriers there. For example, the normal (bcc) phase, $\alpha$-Fe, is stable just below the melting point ($\sim 1530^{\circ}$C), but an additional (fcc) phase, $\gamma$-Fe, is the stable one from $\sim 1380^{\circ}$ to $\sim 910^{\circ}$C. Thus $\alpha$-Fe crystals grown from the melt transform during cooling, and the most direct method for growing single crystals is inaccessible. One widely used remedy for this is to alloy 3%-5% Si with the Fe; this addition stabilizes the $\alpha$-Fe phase up to the melting point, and single crystals can be grown from the alloy melt. However, the crystals obtained by this approach are not of high quality, and the presence of the Si (substitutional for the Fe) decreases the effectiveness of schemes depending on exact cancellation of the electronic scattering. Single crystals of pure iron can be grown by strain-anneal and chemical vapor decomposition techniques, but the crystals obtained are small and of large mosaic spread ($\sim 1^{\circ}$). A logical solution to these problems would be to epitaxially grow a thin film of pure $Fe^{57}$ on a substrate with a small electronic Bragg scattering. It would obviously require a substantial research project to grow high-quality crystals in this way, although there is extensive literature on the epitaxial growth of Fe crystals [5.26].

An alternative approach has been used by the Russian group [5.21], which has chosen to use $\alpha$-$^{57}Fe_2O_3$ as the scattering crystal. Good single crystals of $Fe_2O_3$ are readily grown from borate fluxes without requiring extremely large quantities of the separated isotope. A mosaic of less than 1' is reported [5.21], and very good reduction of the electronic Bragg scattering on the (777) reflection to be used for the nuclear resonance.

The preliminary experiments carried out so far suggest that the following features would be very desirable in an experiment to detect the nuclear Bragg scattering:

1) A grazing incidence mirror, to cut off the SR flux at ~15 keV. This will reduce the background due to scattered radiation and eliminate the presence of λ/n harmonics which might be transmitted by the diffracting crystal.
2) A "pre-monochromator" will reduce the heating of the diffracting crystal by the white radiation beam (now ~1 W; due to increase greatly with high intensity SR sources), reduce the shielding, ozone, and safety problems inherent in having the high intensity white radiation beam in the experimental area, and will reduce the contamination and erosion of the crystal from beam-produced ionization.
3) A timed detector, to enhance the observability of the "slow" nuclear Bragg component over the prompt electronic scattering. Use of a resonant timed detector [5.1] would provide significant discrimination at some cost in counting rate.

Additionally, the complexities and difficulty of the experiment make it necessary to dedicate a beam line for this purpose rather than using a "general purpose" line.

### 5.3.3 Grazing Incidence Reflecting Films – A New Development

Many of the problems discussed above may be solved by the use of "impedance matched grazing incidence films" recently suggested by HANNON et al. [5.27] as shown in Fig.5.7. The basic arrangement involves the use of a thin X-ray reflecting layer on a substrate of different refractive index. By proper choice of $n_1$, $n_2$, $l_f$, and $\phi$, it is possible to make the wave reflected from the surface of the antireflecting layer (ray A in Fig.5.7) of equal amplitude but opposite in phase to the wave (ray B) reflected from the substrate. Thus the two reflections cancel, and the net grazing incidence reflected energy is zero. This arrangement is exactly like the use of antireflective coating on optical surfaces. Thus, it is possible to cancel the X-ray reflection from the electrons in the substrate. At the nuclear resonance energy, however, the resonance produces strong changes in the refractive index of the substrate, leading to new reflectivity terms which are not cancelled by the antireflecting coating. Thus, it should be possible to observe X rays reflected due to the nuclear resonance over a null background of electronically reflected radiation. The structure shown in Fig.5.7 is thus a filter, which passes only X-ray energies in the neighborhood of the nuclear resonance.

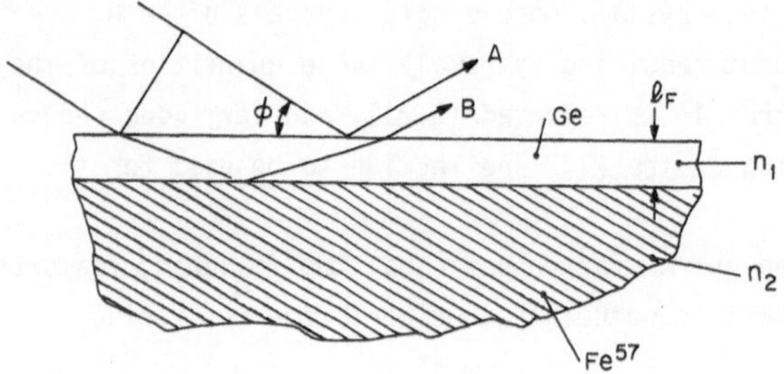


Fig. 5.7. Grazing incidence reflection scheme proposed by HANNON et al. [5.27]. The incidence angle $\phi$ has been greatly enlarged for clarity; values of $\phi \simeq 5 \times 10^{-3}$ rad would be used in an actual experiment

This approach has a number of advantages for studying the interaction of the X-ray beam with the nuclear resonance. Since the electronic parts of $n_1$ and $n_2$ vary only slowly with X-ray energy, cancellation of the electronic reflection term should occur over a relatively wide range of X-ray energy. Neither lattice constants nor the grazing incidence angle $\phi$ need be controlled with the precision required in the various Bragg-scattering cancellation schemes discussed above. Initial calculations [5.27] suggest that attenuations of $10^{-2}$ to $10^{-4}$ can be obtained using, e.g., a Ge anti-reflecting layer of $l_f \simeq 65$ Å, $\phi \simeq 4.6 \times 10^{-3}$ rad, and an Fe substrate, at the 14-keV energy appropriate for studying the $Fe^{57}$ resonance. These calculations assumed perfectly smooth and homogeneous surfaces which are clearly unattainable in practice. On the other hand, it would be easy to pass the beam through several successive reflections from these filters, thus cascading the attenuation.

Preliminary tests have been carried out [5.28] and have already shown attenuation by about a factor of 10. A long development effort to improve surface smoothness and interface sharpness, and reduce granularity and surface contamination, is in prospect, but these early results are very encouraging.

### 5.3.4 Conclusions

Although nuclear Bragg scattering from SR sources has not yet been observed, it is likely that it will be observed within the next year and will rapidly be developed to a stage where useful beams of highly monochromatic radiation are being generated. For the first few years, the primary interest will be in characterizing the details of the nuclear scattering process and the time dependence of the scattered radiation. Eventually, the radiation should be useful for nuclear hyperfine studies and will probably be intense enough to make long base-line interferometers [5.19] and determination of X-ray scattering phase shifts possible. The anticipated increase in intensity of SR sources makes schemes which now appear to be of marginal utility likely to be extremely practical in the long run.

*Acknowledgements*. The author acknowledges with gratitude the permission granted by Plenum Publishing Corporation to reprint in its entirety this chapter which was originally published in *Synchrotron Radiation Research*, edited by Drs. H. Winick and S. Doniach, 1980. This volume contains 22 review articles on various aspects of research using synchrotron radiation.

I thank Dr. Raju Raghavan for suggestions resulting from a reading of the manuscript.

## References

5.1 R.L. Cohen, G.L. Miller, K.W. West: Phys. Rev. Lett. *41*, 381 (1978)
5.2 E.J. Seppi, F. Boehm: Phys. Rev. *128*, 2334 (1962); and references therein

5.3 F.R. Metzger: Prog. Nucl. Phys. *7*, 53 (1959)
5.4 P.J. Black, P.B. Moon: Nature (London) *188*, 481 (1960)
5.5 R.L. Cohen (ed.): *Applications of Mössbauer Spectroscopy* (Academic Press, New York 1976)
5.6 S.L. Ruby: J. Phys. (Paris) *C6*, 209 (1974)
5.7 U. Gonser (ed.): *Mössbauer Spectroscopy*, Topics in Applied Physics, Vol.5 (Springer, Berlin, Heidelberg, New York 1975) pp.43-47
5.8 R.V. Pound, G.A. Rebka, Jr.: Phys. Rev. Lett. *4*, 337 (1960)
5.9 This equivalence has recently been established for microwave frequencies to ~200 ppm by R.C. Vessot: Gen. Relat. Gravit. *10*, 181 (1979)
5.10 P.J. Black, I.P. Duerdoth: Proc. Phys. Soc. (London) *84*, 169 (1964)
5.11 F. Parak, R.L. Mössbauer, U. Biebl, H. Formanek, W. Hoppe: Z. Phys. *244*, 456 (1971)
F. Parak, R.L. Mössbauer, W. Hoppe, U.F. Thomanek, D. Bade: J. Phys. (Paris) *C6*, 703 (1976)
5.12 P.J. Black, G. Longworth, D.A. O'Connor: Proc. Phys. Soc. (London) *83*, 925 (1964)
5.13 This technique has been used for nuclear resonance fluorescence, see F.R. Metzger: Nucl. Phys. *27*, 612 (1961)
5.14 J. Christiansen, H.E. Mahnke, E. Recknagel, D. Riegel, G. Weyer, W. Witthuhn: Phys. Rev. Lett. *21*, 554 (1968)
5.15 E. Matthias: In *Hyperfine Structure and Nuclear Radiations*, ed. by E. Matthias, D.A. Shirley (North Holland, Amsterdam 1968) p.815 ff
5.16 Yu. Kagan, A.M. Afanasev, I.P. Perstenev: Sov. Phys.-JETP *27*, 819 (1968)
5.17 J.P. Hannon, G.T. Trammell: Phys. Rev. *186*, 306 (1969)
5.18 S.L. Ruby: A.I.P. Conf. Proc. No.38, p.50
5.19 G.T. Trammell, J.P. Hannon, S.L. Ruby, P. Flinn, R.L. Mössbauer, F. Parak: A.I.P. Conf. Proc. No.38, p.46, and to be published
5.20 S.L. Ruby, P. Flinn: unpublished
5.21 A.N. Artemyev, V.A. Kabannik, Yu.N. Kazakov, G.N. Kulipanov, E.A. Meleshko, V.V. Sklyarevskiy, A.N. Skrinsky, E.P. Stepanov, V.B. Khlestov, A.I. Chechin: Nucl. Instrum. Methods *152*, 235 (1978)
5.22 R.L. Cohen, P.A. Flinn, E. Gerdau, J.P. Hannon, S.L. Ruby, G.T. Trammell: A.I.P. Conf. Proc. No.38, p.140
5.23 Yu. Kagan, A.M. Afanasev, V.G. Kohn: Phys. Lett. *68*, 339 (1978)
5.24 G.T. Trammell, J.P. Hannon: Phys. Rev. *B18*, 165 (1978)
5.25 E.P. Stepanov, A.N. Artem'ev, I.P. Perstnev, V.V. Sklyarevskii, V.I. Smirnov: Sov. Phys.-JETP *39*, 562 (1974)
5.26 For a bibliography, see E. Grünbaum: In *Epitaxial Growth, Part B* (Academic Press, New York 1975) p.611 ff
5.27 J.P. Hannon, M. Müller, E. Gerdau, H. Winkler, R. Rüffer: Phys. Rev. Lett. *43*, 636 (1979)
5.28 E. Gerdau, J.P. Hannon: private communication

# 6. Resonance $\gamma$-Ray Polarimetry

U. Gonser and H. Fischer

With 19 Figures

"I have at last succeeded in magnetizing and electrifying a ray of light and in illuminating a magnetic line of force." This statement FARADAY wrote proudly after a long struggle to demonstrate an interaction of electric and magnetic forces with light [6.1]. His discovery of the Faraday effect in 1846 opened the door to a new field: magneto- and electro-optics, that is, optical effects involving emitting sources or transmitting media subjected to electric and magnetic fields. Two other important discoveries also go back to the last century: the Kerr effect (1875) as the first electro-optic effect and the Zeeman effect (1896) as the splitting of spectral lines in a magnetic field.

In this century many ideas originating in optics have been applied in the investigation of nuclear $\gamma$ rays, particularly to obtain information regarding energy, intensity, propagation direction, multipolarity, and polarization. The successes, but also the difficulties involved, are described in a paper which was published just at the time when the significance of the Mössbauer effect was being realized [6.2].

The discovery of the Mössbauer effect [6.3] changed the situation drastically. The hyperfine interactions provided relatively simple access to nuclear properties and electron- and spin-density distributions. Of particular interest in Mössbauer spectroscopy are the three hyperfine interactions which correspond to the three principal nuclear moments:

1) electric monopole interaction – isomer shift,
2) magnetic dipole interaction – nuclear Zeeman effect,
3) electric quadrupole interaction – quadrupole splitting.

The latter two interactions remove the energy degeneracies of the nuclear levels so that each line of the emission or absorption spectrum corresponds to transitions between defined nuclear spin states. The intensity of the emitted or absorbed radiation and its dependence on orientation are determined by conservation of angular momentum in the system of nucleus plus $\gamma$ quantum (selection rules).

## 6.1 Intensity and Polarization of Radiation in Mössbauer Transitions

Usually the nuclei of a Mössbauer isotope or its parent isotope are randomly distributed over the sites of the absorber or source material, respectively. Also, there exists no definite relation between the times at which the transitions in different nuclei occur. Thus, the emission and absorption may be considered as incoherent processes[1]. Therefore, in order to find the intensity and polarization of the radiation emitted or absorbed, only transitions in a single nucleus need to be considered. If there are several physically different nuclear sites in a material, the total intensity and polarization are obtained by properly averaging over all different sites. We will treat first just a single nucleus in the excited state and the radiation it emits in decaying to the ground state. Such a transition is characterized by the spins and parities of both levels, and by the multipolarity of the transition. All these quantities have been measured for a large number of isotopes having Mössbauer transitions, and they are given in the literature [6.4].

If neither the ground-state nor the excited-state degeneracy with respect to energy is lifted by a hyperfine interaction, matters are of comparatively little interest since there will be no polarization of the radiation. The radiation is emitted isotropically and the spectral distribution of the intensity is described by a single Lorentzian,

$$I(\omega) = \frac{\Gamma/2\pi}{\hbar(\omega-\omega_0)^2+(\Gamma/2)^2} , \tag{6.1}$$

where $\hbar\omega_0$ is the energy difference between the two states. The half-width $\Gamma$ is determined by the mean lifetime of the excited state. The absolute intensity can hardly be measured and is, at least in principle, of little interest.

In the other case, i.e., when a hyperfine interaction is present, the situation becomes much more complex. Depending on variations of the hyperfine interaction in time, one has to distinguish between several cases. If these variations are very fast as compared to the lifetime of the excited state or to the time needed to assign a definite hyperfine energy to the nuclear hyperfine levels ($t \approx \hbar/\Delta E$, which in the case of the magnetic hyperfine interaction is the Larmor precession time), the nucleus will experience only the mean values of the hyperfine fields during its decay. If these mean values are zero, the situation will be very much like that without any hyperfine fields. This is not true when the variations in time occur at a rate comparable to the nuclear lifetime or precession time. We will ignore this

---

1 For scattering processes in highly enriched materials this is obviously not true. High enrichment gives rise to a correlation of spatial positions of the scatterers. The incident radiation reaches different nuclei with definite delays in time, thus causing a correlation between the times of nuclear transitions (see chapter by Mössbauer, Parak and Hoppe)

possibility and the relaxation phenomena which appear in such cases and restrict ourselves to the simpler case where the hyperfine fields are constant during the decay of the nucleus.

In the latter case the spectrum consists of several lines of different energy. By summing up all lines one again obtains isotropic intensity and no polarization. But with the high resolution of Mössbauer spectroscopy, it is possible to observe the hyperfine lines separately, and their intensities are both angle dependent and polarized. The polarization may be complete or partial depending on the kind of hyperfine interaction and also on the direction of observation relative to the hyperfine fields.

### 6.1.1 Description of Polarization in Terms of Density Matrices

The quantum-mechanical treatment of electromagnetic radiation leads to the introduction of photons which are bosons of vanishing rest mass. An important property of photons [6.5] is that regardless of the multipolarity of the radiation determined by the angular momentum L (L>0), the helicity of a photon, i.e., the projection of its angular momentum onto the direction of propagation, may only have the values ±1. It might be helpful to note that for photons, as well as any other particles with zero rest mass, the separation of total angular momentum into spin and orbital angular momentum does not really make sense since there exists no rest frame. The two helicity eigenstates |+> and |-> therefore constitute a very convenient set of basic states for the representation of the photon angular momentum state ${}^{\pi}\chi_L^M$, where the index $\pi$ refers to the parity of the radiation and L and M are the J and $J_z$ quantum numbers, respectively.

For Mössbauer spectroscopy only the two lowest multipolarities, i.e., dipole (L=1) and quadrupole radiation (L=2), are of importance. The helicity representation of photon angular momentum states for these two multipolarities is given in Table 6.1.

By choosing a coordinate system with the $\gamma$-radiation propagation direction along z, the helicity representation simplifies to

$$\chi_1^1 = \mp\sqrt{\frac{3}{8\pi}}\,|+\rangle \qquad \chi_1^{-1} = \sqrt{\frac{3}{8\pi}}\,|-\rangle \tag{6.2a}$$

for dipole radiation, and

$$\chi_2^1 = \pm\sqrt{\frac{5}{8\pi}}\,|+\rangle \qquad \chi_2^{-1} = \sqrt{\frac{5}{8\pi}}\,|-\rangle$$

for quadrupole radiation. The upper and lower signs correspond to even and odd parities, respectively. The normalization of these states is such that

$$\int \chi_L^{*M}\,\chi_{L'}^{M'}\,d\Omega = \delta_{MM'}\delta_{LL'} \quad . \tag{6.2b}$$

Table 6.1. Helicity representation of photon states $\chi_L^M$ [6.6]. $\vartheta$, $\varphi$ are the polar and azimuthal angles of the propagation direction in a system where L and M are the $\hat{J}$ and $\hat{J}_z$ quantum numbers, respectively

(a) Dipole. The upper signs denote E1 radiation, the lower signs denote M1 radiation

$$\chi_1^1 = \sqrt{\frac{3}{16\pi}}\left(\pm\frac{1+\cos\theta}{\sqrt{2}}|+> + \frac{1-\cos\theta}{\sqrt{2}}|->\right)e^{i\varphi}$$

$$\chi_1^0 = \sqrt{\frac{3}{16\pi}}\sin\theta(\pm|+>-|->)$$

$$\chi_1^{-1} = \sqrt{\frac{3}{16\pi}}\left(\pm\frac{1-\cos\theta}{\sqrt{2}}|+> + \frac{1+\cos\theta}{\sqrt{2}}|->\right)e^{-i\varphi}$$

(b) Quadrupole. The upper signs denote E2 radiation, the lower signs denote M2 radiation

$$\chi_2^2 = \sqrt{\frac{5}{16\pi}}\sin\theta\left(\mp\frac{1+\cos\theta}{\sqrt{2}}|+> - \frac{1-\cos\theta}{\sqrt{2}}|->\right)e^{2i\varphi}$$

$$\chi_2^1 = \sqrt{\frac{5}{16\pi}}\left(\pm\frac{2\cos^2\theta+\cos\theta-1}{\sqrt{2}}|+> - \frac{2\cos^2\theta-\cos\theta-1}{\sqrt{2}}|->\right)e^{i\varphi}$$

$$\chi_2^0 = \sqrt{\frac{5}{16\pi}}\sqrt{3}\sin\theta\cos\theta(\pm|+>-|->)$$

$$\chi_2^{-1} = \sqrt{\frac{5}{16\pi}}\left(\mp\frac{2\cos^2\theta-\cos\theta-1}{\sqrt{2}}|+> + \frac{2\cos^2\theta+\cos\theta-1}{\sqrt{2}}|->\right)e^{-i\varphi}$$

$$\chi_2^{-2} = \sqrt{\frac{5}{16\pi}}\sin\theta\left(\pm\frac{1-\cos\theta}{\sqrt{2}}|+> + \frac{1+\cos\theta}{\sqrt{2}}|->\right)e^{-2i\varphi}$$

The states with M = +1, -1 correspond to right- and left-circular polarization. Any other pure state of polarization may be written as a linear combination, corresponding to a superposition, of these basis states:

$$|\psi_P> = \sum_L \frac{a_L}{\sqrt{2L+1}} \sum_{M=-1,1} \varepsilon_L^M \chi_L^M \quad , \tag{6.3}$$

where the summation over different L takes into account the possibility of a multipole mixture. As a consequence of time-reversal symmetry, the coefficients $a_L$ may be chosen to be real, and their ratios $|a_L/a_{L'}|^2$ determine the multipole mixture. Thus, any state of pure polarization may be represented by a two-dimensional state vector in the helicity representation. This is not true for a state of partial polarization which can be regarded as an incoherent superposition of states with pure polarization. In principle, it would be possible to describe such general states of polarization by specifying each time the superposition of states is coherent or incoherent, i.e., whether amplitudes or their squares have to be added. But this would be a rather tedious and inconvenient procedure. However, the formalism of density matrices offers a convenient method of treating such incoherent superpositions formally [6.7].

The density operator of a pure state $|\psi_P\rangle$ may be defined as just the outer product

$$\hat{\rho}_P = |\psi_P\rangle\langle\psi_P| \tag{6.4}$$

with the effect that the overall phase of this state is cancelled. The density operator of a mixed state, being the incoherent superposition of the pure states $|\psi_{P1}\rangle$ and $|\psi_{P2}\rangle$, is obtained by

$$\hat{\rho} = \hat{\rho}_1 + \hat{\rho}_2 = |\psi_{P1}\rangle\langle\psi_{P1}| + |\psi_{P2}\rangle\langle\psi_{P2}| \quad . \tag{6.5}$$

This density operator still contains all information of physical significance. From its definition it can be seen that $\hat{\rho}$ is Hermitian and that the intensity for a pure state $I = \langle\psi_P|\psi_P\rangle$ is given by the trace

$$I = \mathrm{Tr}\{\hat{\rho}\} \quad , \tag{6.6}$$

which for the general case reflects the rule of adding intensities instead of amplitudes when dealing with incoherent superpositions.

In any representation, the density matrix of the unpolarized state is proportional to the unit matrix $\hat{E}$, and by use of

$$\hat{\rho} = \lambda\hat{E} + (\hat{\rho}-\lambda\hat{E}) \tag{6.7}$$

the density matrix of a partially polarized state may be decomposed into the incoherent superposition of the unpolarized state and a completely polarized state. Since the determinant of the density matrix of a pure state vanishes, $\lambda$ is obtained from

$$\det\{\hat{\rho}-\lambda\hat{E}\} = 0 \quad . \tag{6.8}$$

Of the two solutions for $\lambda$, only the one which fulfills $\mathrm{Tr}(\lambda\hat{E}) \geq 0$, $\mathrm{Tr}(\hat{\rho}-\lambda\hat{E}) \geq 0$ is meaningful, leading to the positive degree of polarization

$$\xi = \sqrt{1-4\,\frac{\det\hat{\rho}}{(\mathrm{Tr}\{\hat{\rho}\})^2}} \qquad 0 \leq \xi \leq 1 \quad . \tag{6.9}$$

Another way of visualizing the polarization represented by a given density matrix is to decompose it into a pair of density matrices describing pure states, i.e., to regard it as the incoherent superposition of two orthogonal polarization states. Such a decomposition is given by

$$\hat{\rho} = \lambda_1|\psi_1\rangle\langle\psi_1| + \lambda_2|\psi_2\rangle\langle\psi_2| \quad , \tag{6.10}$$

where $\lambda_1$, $\lambda_2$ are the real eigenvalues of $\hat{\rho}$ and $|\psi_1\rangle$, $|\psi_2\rangle$ are the corresponding eigenvectors. Since $\hat{\rho}$ is Hermitian, the only case where this decomposition is not unique is that of the unpolarized state, which can be thought of as the incoherent superposition of any two orthogonal states.

### 6.1.2 Calculation of Density Matrices for the Hyperfine Lines

As already mentioned, the radiation emitted by a decaying nucleus may be polarized only if there is a hyperfine interaction which at least partially lifts the degeneracy of the nuclear levels. In order to calculate the polarization of a particular hyperfine line, one has to know the two nuclear states (initial and final) of the corresponding transition. In the general case each of these states may be degenerate, i.e., there may be several orthogonal state vectors spanning the spaces of the two hyperfine levels involved in the transition under consideration. Again, the formalism of density matrices allows for a convenient description of such states. The density matrix of an excited hyperfine level reads

$$\hat{\rho}_e = \frac{1}{2I_e+1} \sum_i |e^i\rangle\langle e^i| \quad , \tag{6.11}$$

where the $\{|e^i\rangle\}$ constitute an orthogonal set of nuclear states all having the energy of the given hyperfine level. It has been assumed that the states $|e^i\rangle$ are equally populated. (Due to the previous history of the nucleus, however, this may not necessarily be true.) Analogously the ground hyperfine level is described by

$$\hat{\rho}_g = \frac{1}{I_g+1} \sum_f |g^f\rangle\langle g^f| \quad . \tag{6.12}$$

The excited state may now be regarded as a composite system consisting of the ground state and the radiation. That is,

$$\hat{\rho}_e = \hat{\rho}_r \times \hat{\rho}_g \tag{6.13}$$

and the density matrix of the radiation $\hat{\rho}_r$ are obtained from

$$\hat{\rho}_r = \mathrm{Tr}_g(\hat{\rho}_r \times \hat{\rho}_g) \quad , \tag{6.14}$$

where $\mathrm{Tr}_g$ denotes the trace over all ground-state variables. The state vectors for the nuclear states may be written as

$$|e^i\rangle = \sum_{\overline{m}=-I_e}^{I_e} \varepsilon^i_{\overline{m}} |I_e \overline{m}\rangle$$

$$|g^f\rangle = \sum_{\underline{m}=-I_g}^{I_g} \gamma^f_{\underline{m}} |I_g \underline{m}\rangle \quad , \tag{6.15}$$

where $|Im\rangle$ are the spin states for the excited state ($I=I_e$) and the ground state ($I=I_g$), respectively. The decomposition of the excited-state vectors into ground-state vectors and radiation states is accomplished by using

$$|I_e\overline{m}\rangle = \sum_L \frac{a_L}{\sqrt{2L+1}} \sum_{M,\underline{m}} C^{I_e\, L\, I_g}_{\overline{m}\; M\, \underline{m}}\, \chi^M_L\, |I_g\underline{m}\rangle \tag{6.16}$$

according to the quantum-mechanical rule for the coupling of angular momenta. The $\chi^M_L$ are the photon spin states, the $C^{I_3 I_2 I_1}_{m_3 m_2 m_1}$ are the Clebsch-Gordan coefficients defined by

$$C^{I_3 I_2 I_1}_{m_3 m_2 m_1} = \langle I_3 m_3 | I_2 m_2, I_1 m_1 \rangle \tag{6.17}$$

and the $a_L$ determine the multipole mixing of the transition.

By following the procedure outlined above, one obtains as the result of some algebra the density matrix of the radiation

$$\hat{\rho}_r = \sum_{i,f} \sum_{L,L'} \frac{a_L a^*_{L'}}{\sqrt{(2L+1)(2L'+1)}} \sum_{M,M'} \sum_{\underline{m},\underline{m}'} \varepsilon^i_{(\underline{m}+M)} \varepsilon^{i*}_{(\underline{m}'+M')} \times \gamma^f_{\underline{m}'} \gamma^{f*}_{\underline{m}}\, C^{I_e\quad\; L\, I_g}_{(\underline{m}+M)\, M\, \underline{m}}\, C^{I_e\quad\;\; L'\, I_g}_{(\underline{m}'+M')\, M'\, \underline{m}'}\, \chi^M_L\, \chi^{M'*}_{L'} \quad , \tag{6.18}$$

where the summation has already been simplified by application of the principle of conservation of z-components of angular momenta. Since the photon states are not linearly independent in the helicity representation (see Table 6.1), there are also terms due to interference between the contributions of different multipole order which are reflected in the density matrix and can therefore also be observed experimentally in the line intensities and polarizations.

As can be seen from Table 6.1, the helicity representation of the photon states becomes rather simple in a coordinate system with z-axis parallel to the propagation direction. In such a system only the values $M = \pm 1$ contribute to the density matrix. We will refer to this system as the laboratory frame. Usually the coefficients in (6.18) are most easily calculated in a system related to the hyperfine fields causing the splitting. When this system with its coefficients $\varepsilon'$, $\gamma'$ is obtained from the laboratory frame by the Euler rotation $\omega = (\varphi, \vartheta, \psi)$, the coefficients $\varepsilon$ and $\gamma$ in the laboratory system are given by

$$\varepsilon^i_{\overline{m}} = \sum_{\overline{m}'} \varepsilon'^i_{\overline{m}'}\, \mathcal{D}^{(I_e)}_{\overline{m}'\overline{m}}\, (\omega^{-1})$$
$$\gamma^f_{\underline{m}} = \sum_{\underline{m}'} \gamma'^f_{\underline{m}'}\, \mathcal{D}^{(I_g)}_{\underline{m}'\underline{m}}\, (\omega^{-1}) \quad , \tag{6.19}$$

where the

$$\mathcal{D}^{(I)}_{m'm}(\varphi,\vartheta,\psi) = e^{im'\varphi}\, d^{(I)}_{m'm}\, (\vartheta)\, e^{im\psi} \tag{6.20}$$

are matrix elements of the (2I+1)-dimensional irreducible representation of the rotation group. In (6.19) the inverse rotation,

$$\omega^{-1} = (\pi-\psi, \vartheta, \pi-\varphi) \quad , \tag{6.21}$$

has to be used.

Taking into account the energy dependence of the lines in the spectrum, the total density matrix of the radiation emitted (or absorbed) by one nucleus reads

$$\hat{\rho}(\omega) = \sum_{\ell} \mathrm{Tr}\, \hat{\rho}_{\ell} \frac{\hat{\rho}_{\ell}}{\mathrm{Tr}\, \hat{\rho}_{\ell}} \frac{\Gamma/2\pi}{\hbar(\omega-\omega_{\ell})^2+(\Gamma/2)^2} \quad , \tag{6.22}$$

where the density matrices of all lines have been normalized to trace one.

When no multipole mixing is present, the intensities for transitions between pure m-states of the nucleus as given by Tr $\hat{\rho}_r$ reduce to

$$I(\vartheta, I_e, m_e, I_g, m_g) \propto \begin{bmatrix} I_e & L & I_g \\ \zeta m_e & M & m_g \end{bmatrix}^2 \left\{ d_{1M}^{(L)^2} + d_{-1M}^{(L)^2} \right\} \quad . \tag{6.23}$$

The essential difference between the angular dependences of dipole (L=1) and quadrupole (L=2) transitions may be seen from Figs.6.1,2 which show the intensities of a $I_e = 3/2 \leftrightarrow I_g = 1/2$ dipole transition in contrast to those of the corresponding quadru-

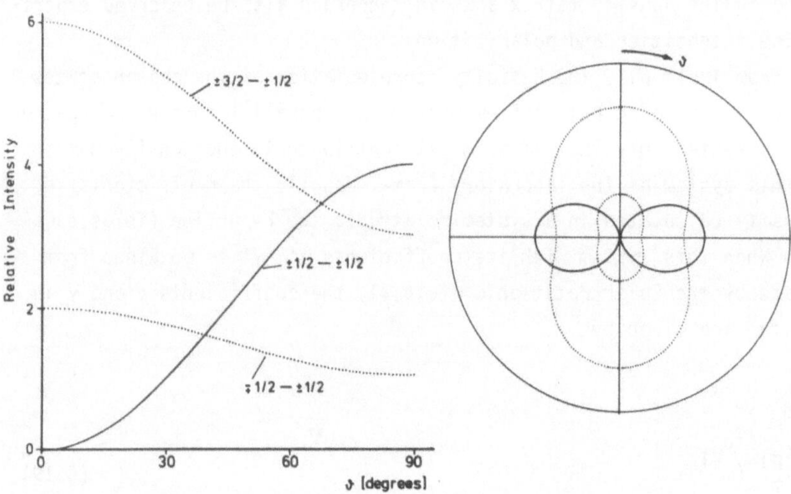


Fig. 6.1. Relative line intensities of the allowed dipole transitions between states with spins 3/2 and 1/2 as a function of the angle ϑ between the γ radiation and the orientation of the magnetic field. The sum of the contributions yields isotropic intensities as indicated by the circle

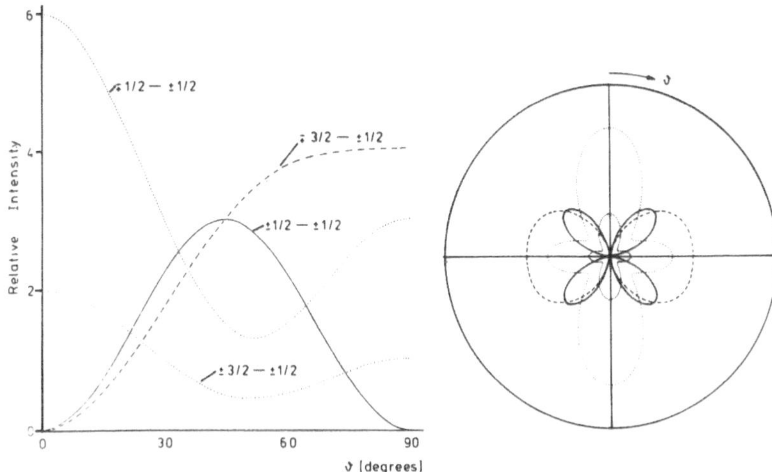


Fig. 6.2. Relative line intensities of the allowed quadrupole transitions between states with spins 3/2 and 1/2 as a function of the angle $\vartheta$ between the $\gamma$-ray direction and the orientation of the magnetic field. The sum of the contributions yields isotropic intensities as indicated by the circle

pole transition. Both types of transitions have been observed experimentally; the dipole case is represented by a variety of isotopes including the commonly used $^{57}$Fe and $^{119}$Sn, whereas the quadrupole case has been verified for $^{201}$Hg. The other possible transitions of the available isotopes are listed in the Mössbauer Effect Data Index [6.4].

### 6.1.3 Transmission of $\gamma$ Rays in Resonantly Absorbing Materials

When $\gamma$ radiation passes through a medium, it is attenuated by absorption. There are several different absorption processes which may be separated roughly into energy-dependent and energy-independent types. For Mössbauer spectroscopy the interesting range of energies is confined to the vicinity of the resonance energy. Within this range the only absorption process showing considerable energy dependence is resonant absorption by the appropriate nuclei. An energy-dependent absorption always gives rise to an energy-dependent dispersion, and these two may be combined into a complex index of refraction $\hat{n}$. The imaginary part of $\hat{n}$ is related to the absorption and the real part to the dispersion. The two are related to each other by dispersion relations. Thus, knowledge of the absorption readily gives the dispersion and vice versa. The resonant absorption depends on polarization and is described for a given direction by a 2×2 density matrix. Therefore, the index of refraction must also depend on polarization and be a complex 2×2 matrix.

In general there may be different lattice sites for the Mössbauer isotopes labelled by the index j, and for each of these sites, there may be several transitions due to the hyperfine splitting. The density matrix for transition i of a nucleus in site j is $\hat{\rho}_j^i$. With $N_j$ nuclei per unit volume in site j and recoil-free fractions $f_j$, the index of refraction for an energy E comparable to the resonance energies is given by [6.8-10]

$$\hat{n} = \hat{E} - \frac{\sigma_0}{2k_0} \sum_j N_j f_j \sum_i \hat{\rho}_j^i \frac{\Gamma/2}{E-E_{ij}+i(\Gamma/2)} \quad ; \tag{6.24}$$

where the density matrices are normalized according to $\sum_i \hat{\rho}_j^i = \hat{E}$, the unit matrix; $\sigma_0$ is the nuclear cross section at resonance; and $k_0 = E/\hbar c$. The energy- and polarization-independent nonresonant absorption may be taken into account by adding a constant times the unit matrix to the imaginary part of $\hat{n}$. The density matrix of radiation which has traversed a distance d in the absorber will have changed to

$$\hat{\rho}(d) = e^{i\hat{n}kd}\, \hat{\rho}_0\, e^{-i\hat{n}^+kd} \quad . \tag{6.25}$$

The matrices $\hat{\rho}_0$ and $\hat{n}kd$ may be reexpressed in terms of the unit matrix and the three Pauli matrices [6.10]

$$\hat{\sigma}_\xi = \begin{pmatrix} 0 & 1 \\ 1 & 0 \end{pmatrix} \quad , \quad \hat{\sigma}_\eta = \begin{pmatrix} 0 & -i \\ i & 0 \end{pmatrix} \quad , \quad \hat{\sigma}_\zeta = \begin{pmatrix} 1 & 0 \\ 0 & -1 \end{pmatrix} \quad ,$$

by writing them in the following forms:

$$\hat{\rho}_0 = \frac{1}{2} I_0(\hat{E} + \underline{P} \cdot \underline{\sigma}) \tag{6.26a}$$

$$\hat{n}kd = a \cdot \hat{E} + \underline{b} \cdot \underline{\sigma} \quad , \tag{6.26b}$$

where

$$I_0 \equiv \mathrm{Tr}\, \hat{\rho}_0 \quad , \quad P_\mu \equiv \mathrm{Tr}(\hat{\rho}_0 \cdot \hat{\sigma}_\mu) \quad , \quad a \equiv \frac{1}{2} kd\, \mathrm{Tr}\, \hat{n} \quad , \quad b_\mu \equiv \frac{1}{2} kd\, \mathrm{Tr}\, (\hat{n} \cdot \hat{\sigma}_\mu)$$

with

$$\mu = \xi, \eta, \zeta \quad .$$

Since the intensity is given by

$$I(d) = \mathrm{Tr}\{\hat{\rho}(d)\} = \mathrm{Tr}\left\{e^{i\hat{n}kd}\, \hat{\rho}_0\, e^{-i\hat{n}^+kd}\right\} \quad , \tag{6.26c}$$

we are now in a position to represent polarized radiation in terms of an intensity I and a polarization vector $P = (P_\xi, P_\eta, P_\zeta)$, both of which depend on the thickness and the optical properties of the absorbing medium and on the intensity and polarization of the incident radiation. This is called the Poincaré representation.

Substituting (6.26a) and (6.26b) into (6.26c) gives for the transmitted intensity in the Poincaré representation

$$I(d) = I_0 \exp[i(a-a^*)][\cos b^* \cos b + (\hat{b}^* \cdot \hat{b}) \sin b^* \sin b - i(\hat{b}^* \cdot \underline{P}) \cdot \sin b^* \cos b + i(\hat{b} \cdot \underline{P}) \sin b \cos b^* + i\underline{P}(\hat{b}^* \times \hat{b}) \sin b^* \sin b] \tag{6.27a}$$

$$\underline{P}(d) \cdot I(d) = I_0 \exp[i(a-a^*)]\Big\{i\hat{b} \sin b \cos b^* - i\hat{b}^* \sin b^* \cos b - i(\hat{b}^* \times \hat{b}) \sin b^* \sin b + \underline{P} \cos b^* \cos b + (\underline{P} \times \hat{b}) \sin b \cos b^* + (\underline{P} \times \hat{b}^*) \sin b^* \cos b + [\hat{b} \cdot (\underline{P} \cdot \hat{b}^*) + \hat{b}^* (\underline{P} \cdot \hat{b}) - \underline{P}(\hat{b} \cdot \hat{b}^*)] \sin b^* \sin b\Big\} , \tag{6.27b}$$

where $b = (b_\xi^2 + b_\eta^2 + b_\zeta^2)^{\frac{1}{2}}$ and $\hat{b} = \underline{b}/b$.

In order to get a clearer impression of how the intensity and polarization are affected by transmission through a resonant absorber, some special cases will be considered in more detail.

*a) Thin Absorber Approximation*

By setting $\tau_j = \sigma_0 N_j f_j d$, a dimensionless effective thickness is defined. The expansion of (6.25) up to terms linear in d yields

$$\hat{\rho}(d) \approx \hat{\rho}_0 - \frac{i}{2} \sum_j \tau_j \sum_i \left( \hat{\rho}^i_j \hat{\rho}_0 \frac{\Gamma/2}{E-E_{ij}+i(\Gamma/2)} - \hat{\rho}_0 \hat{\rho}^i_j \frac{\Gamma/2}{E-E_{ij}-i(\Gamma/2)} \right) . \tag{6.28}$$

The intensity of the transmitted radiation is then

$$I(d) \approx I_0 - \sum_j \tau_j \sum_i \mathrm{Tr}\left\{\hat{\rho}^i_j \hat{\rho}_0\right\} \frac{\Gamma/2}{(E-E_{ij})+(\Gamma/2)^2} . \tag{6.29}$$

Thus, the relative intensity absorbed in thin absorbers ($\tau_j \ll 1$) by transition i(j) is proportional to $\mathrm{Tr}\left\{\hat{\rho}^i_j \hat{\rho}_0\right\}$, a result which is frequently used in the analysis of hyperfine field orientations with polarized $\gamma$ rays.

*b) Independent or Well-Separated Absorber Transitions*

When $\hat{n}$ commutes with its adjoint $\hat{n}^+$, the calculation of intensities is simplified considerably (since then $\exp(-i\hat{n}^+ kd)\exp(i\hat{n}kd) = \exp[-ikd(\hat{n}^+ - \hat{n})]$, and $ikd(\hat{n}^+ - \hat{n})$ is Hermitian). Thus, it has real eigenvalues and is diagonal in the representation of its eigenstates. These eigenstates may then be used as a basis for representation of the density matrix so that the intensity becomes

$$I(d) = \mathrm{Tr}\left\{\hat{\rho}_0 \, e^{-ikd(\hat{n}^+ - \hat{n})}\right\} = \sum_{i=1,-1} \rho_{ii} \, e^{-\nu_i} , \tag{6.30}$$

where the $\rho_{ii}$ are the diagonal elements of $\hat{\rho}_0$ in the eigenrepresentation of $ikd(\hat{n}^+ - \hat{n})$ and the $\nu_i$ are the eigenvalues of $ikd(\hat{n}^+ - \hat{n})$.

If $\hat{\rho}(d)$ is diagonal in this representation, i.e., if the incoming radiation is an incoherent superposition of the eigenstates of $ikd(\hat{n}^+ - \hat{n})$, then each eigenstate is only attenuated by the factor $e^{-\nu_i}$, but otherwise transmitted without change (no "rotation"). Since the density matrix of unpolarized radiation is proportional to the unit matrix in any representation, the outcoming radiation must also be an incoherent superposition of the eigenstates of $ikd(\hat{n}^+ - \hat{n})$

$$\hat{\rho}(d) = I_0 \sum_i |i\rangle\langle i| \, e^{-\nu_i} \quad . \tag{6.31}$$

This provides a convenient way of polarizing an initially unpolarized beam by transmitting it through a resonant absorber (filter). The efficiency of such a filter is given by the differences in the eigenvalues $\nu_i$.

The condition that $\hat{n}$ commutes with $\hat{n}^+$ may be put in the form

$$kd[\hat{n},\hat{n}^+] = \frac{1}{4} \sum_{j,j'} \tau_j \tau_{j'} \sum_{i,i'} \left[\hat{\rho}^i_j, \hat{\rho}^{i'}_{j'}\right] \frac{\Gamma/2}{E-E_{ij}+i(\Gamma/2)} \, \frac{\Gamma/2}{E-E_{ij}-i(\Gamma/2)} \tag{6.32}$$

from which it can be seen that a good polarizing filter must fulfill either:

1) all absorption matrices commute with each other or
2) all absorption lines are well separated in energy.

Note that condition 1) is always fulfilled for pure quadrupole splitting in a single crystal with one unique lattice site for the $^{57}$Fe atoms.

## 6.2 Hyperfine Interactions in $^{57}$Fe

The story of the dramatic discovery of the polarization of $^{57}$Fe $\gamma$ radiation in the early days of Mössbauer spectroscopy is told by Stan Hanna in the last chapter.

There are two kinds of hyperfine interactions which might cause an energy splitting of the nuclear states and thus lead to several polarized lines, namely the magnetic dipole and the electric quadrupole interaction of the nucleus with its surroundings. The magnitudes of the splittings reveal the values of the internal magnetic field ($H_{int}$) or the electric field gradient (EFG), respectively. The intensity and orientation dependence of the emitted (or absorbed) radiation are determined by the conservation of angular momentum in the system of nucleus plus $\gamma$ quantum (selection rules). Here we want to restrict ourselves to magnetic dipole radiation and particularly to the 14.4-keV $\gamma$ ray of the first excited state of $^{57}$Fe, because most of the polarization experiments have been performed with this isotope. The energies of the nuclear sublevels for magnetic hyperfine splitting (nuclear Zeeman effect) and quadrupole splitting of the ground and first excited states of $^{57}$Fe are shown in Fig.6.3. The multipolarity of the 14.4-keV $\gamma$ rays is almost ex-

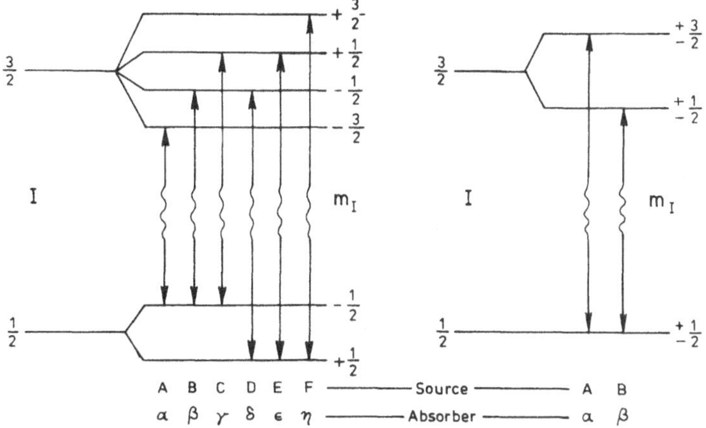


Fig. 6.3. $^{57}$Fe nuclear level diagram for magnetic dipole and electric quadrupole interactions. See text for explanation of the Roman and Greek letters at the bottom of the figure

clusively M1 with the selection rule $\Delta m = 0, \pm 1$; the allowed transitions leading to the six-line Zeeman pattern and to the two-line quadrupole pattern are indicated in the figure. The following notation is adopted here: the Zeeman and quadrupole-split emission lines are designated by the Roman letters A, B, C, D, E, F and A, B, respectively, and the Zeeman and quadrupole split absorption lines by the Greek letters $\alpha$, $\beta$, $\gamma$, $\delta$, $\varepsilon$, $\eta$ and $\alpha$, $\beta$, respectively. In the case of quadrupole splittings, A,$\alpha$ corresponds to the transition $|\pm 3/2\rangle$ and B,$\beta$ to $|\pm 1/2\rangle$. Furthermore, the subscripts S and A, (e.g., $H_S$, $H_A$) relate to source and absorber, respectively.

In order to apply the formulas of Sect.6.1, one has to find the nuclear eigenstates of these interactions. In the general case both kinds of hyperfine fields may be present and the Hamiltonian then is given by

$$H = \frac{eQV_{zz}}{4I(2I-1)} \left[3I_z^2 - I(I+1) + \eta(I_x^2 - I_y^2)\right] - \mu_I(I_xH_x + I_yH_y + I_zH_z) \quad , \tag{6.33}$$

where the magnetic field is described by its components $H = (H_x, H_y, H_z)$, the electric quadrupole field by $V_{zz} = eq$, and the asymmetry parameter

$$\eta = \frac{V_{xx} - V_{yy}}{V_{zz}}$$

with $|V_{zz}| \geqq |V_{yy}| \geqq |V_{xx}|$; $V_{zz} + V_{yy} + V_{xx} = 0$, and thus $0 \leqq \eta \leqq 1$. I = 3/2 and 1/2 for the excited and ground state, respectively, and the $\mu_I$ are the magnetic moments of these states. The ground state (I=1/2) does not have a quadrupole moment, whereas that of the excited state is denoted by eQ. If one of the hyperfine fields strongly dominates the other, approximate analytical solutions of $H$ may be obtained by application of first- or second-order perturbation theory. In the general case of mixed hyperfine interactions, however, the eigenvalues and eigenstates of $H$ for the excited state (I=3/2) have to be calculated numerically.

### 6.2.1 Magnetic Dipole Interaction

In a magnetic hyperfine field the energy degeneracies of both nuclear levels is lifted completely, and thus each line of the emission or absorption spectrum corresponds to transitions between defined nuclear spin states [6.11-14]. The eigenvalues are $E_i = m_I \mu_I H$ with the corresponding eigenstates $|Im\rangle$ in a coordinate system with quantization axis along H.

The six allowed transitions (Fig.6.3) show only three different intensities, and also only three different states of polarization are present corresponding to the three values for $\Delta m = 0, \pm 1$. As a result lines B(β) and E(ε) (Δm=0) are equal in intensity and polarization. Lines A(α) and D(δ) (Δm=+1) have equal polarization and an intensity ratio, $I_{A(\alpha)}:I_{D(\delta)} = 3:1$. The same holds for lines C(γ) and F(η) which are equal in intensity to lines D(δ) and A(α), respectively, but are polarized with opposite helicity (Δm=-1). Taking note of these properties, the density matrices for lines A, ..., F and α, ..., η can be obtained as

$$\hat{\rho}_{B,\beta} = \hat{\rho}_{E,\varepsilon} = \frac{1}{8}\sin^2\vartheta \begin{pmatrix} 1 & e^{2i\varphi} \\ e^{-2i\varphi} & 1 \end{pmatrix}$$

$$\hat{\rho}_{A,\alpha} = 3\hat{\rho}_{D,\delta} = 3\hat{\rho}^{x}_{C,\gamma} = \hat{\rho}^{x}_{F,\eta} = \frac{3}{32}\begin{pmatrix} (1+\cos\vartheta)^2 & -\sin^2\vartheta\, e^{2i\varphi} \\ -\sin^2\vartheta\, e^{-2i\varphi} & (1-\cos\vartheta)^2 \end{pmatrix} , \qquad (6.34)$$

where $\vartheta$ and $\varphi$ are the polar and azimuthal angles, respectively, for the direction of the magnetic hyperfine field. The polar axis is parallel to the γ-ray direction. The elements of matrix $\hat{\rho}^x$ are related to the elements of $\hat{\rho}$ by $\rho^x_{M,M'} = \rho^*_{-M,-M'}$ in the helicity representation where M and M' refer to the basis states $|+\rangle$ and $|-\rangle$, respectively.

Lines B and E (β and ε) are always linearly polarized, whereas the polarization of the other lines ranges from circular (radiation emitted parallel to H) through elliptical to linear (radiation emitted perpendicular to H). When all lines are linearly polarized (radiation emitted perpendicular to H), the polarization of lines B and E (β and ε) is perpendicular to that of lines A, C, D and F (α, γ, δ and η). The sum of all density matrices is half the unit matrix.

From (6.34) it is clear that the intensities of lines A and F (α and η) are equal and are always three times the (equal) intensities of lines C and D (γ and δ). The intensities of the B and E (β and ε) lines related to the others depends on the value of $\vartheta_m$, the angle between the direction of electron spin alignment and the γ-ray direction. The ratio of the intensity of the B or E (β or ε) line, which is a $\pm 1/2 \rightarrow \pm 1/2$ transition with $\Delta m = 0$, to the intensity of the C or D (γ or δ) line, which is a $\pm 1/2 \rightarrow \mp 1/2$ transition with $\Delta m = \pm 1$, is denoted by $R_m$, which turns out to be

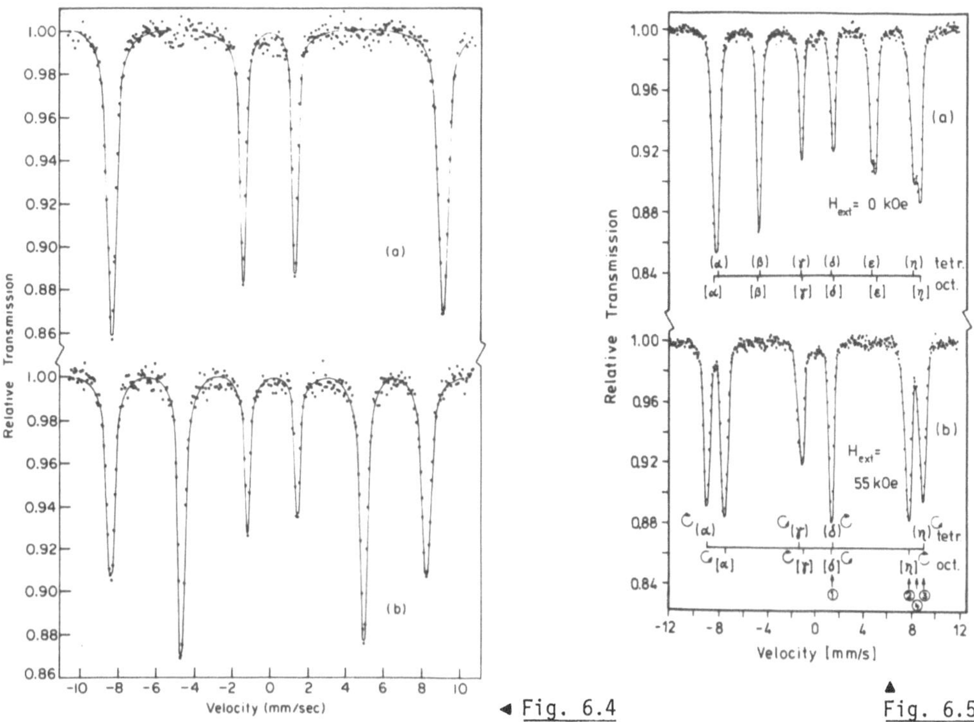


Fig. 6.4a,b. Spectra from a single crystal of $\alpha$-$Fe_2O_3$ cut parallel to the basal plane and with the $\gamma$-ray direction along <111>: (a) at 80 K and (b) at 300 K. A source of $^{57}Co$ in Pt was used

Fig. 6.5a,b. Spectra obtained with a $^{57}Co$-in-Pt source at room temperature and an absorber of polycrystalline ferrimagnetic $(Mg_{0.26},Fe_{0.74})$ $[Mg_{0.74},Fe_{1.26}]O_4$ at 11 K: (a) no field applied, (b) in a magnetic field $H_{ext}$ = 55 kOe parallel to the $\gamma$-ray direction. The positions of the lines corresponding to the tetrahedral (Fe) and octahedral [Fe] sublattices are indicated. See text for explanation of symbols ①, ②, ③, and ④

$$R_m = \frac{4\sin^2\vartheta_m}{1+\cos^2\vartheta_m} \quad . \tag{6.35}$$

When the Debye-Waller factor is direction-independent (no Goldanskii-Karyagin effect) and the electron spin directions are randomly oriented, the average value of $R_m$ is 2. The same value of $R_m$ also occurs for the unique angle $\vartheta_m = 54^\circ 44'$.

First, experiments are discussed where either the source or the absorber exhibits hyperfine splittings while the other consists of a single line (nonpolarized).

The spectra in Fig.6.4 were obtained with a single crystal of $\alpha$-$Fe_2O_3$ (hematite). The crystal was cut parallel to the basal plane and measured (a) at 80 K and (b) at room temperature. The change in the relative line intensities indicates a reorientation of the spins (Morin transition). Below the Morin temperature ($T_M \approx 260$ K) the spins are oriented perpendicular to the basal plane of the rhombohedral structure and parallel or antiparallel to the $\gamma$-ray direction. Thus, the $\Delta m = 0$ lines disappear.

Above the Morin temperature the spins flip and align into the basal plane and the $\Delta m = 0$ lines become strong.

In ferrimagnets the sublattices are usually antiferromagnetically coupled and the net moment results from an imbalance in the magnitudes of magnetizations of the sublattices. Therefore, the net moment is parallel to the magnetization of the dominant sublattice. In the partially inverse spinel $MgFe_2O_4$, the octahedral site [Fe] is always more highly populated than the tetrahedral site (Fe): $(Mg_x, Fe_{1-x})[Mg_{1-x}, Fe_{1+x}]O_4$. In Fig.6.5a the spectrum for the polycrystalline spinel $(Mg_{0.26}, Fe_{0.74})[Mg_{0.74}, Fe_{1.26}]O_4$ is shown [6.15]. The stick diagram indicates the line positions for the tetrahedral (Fe) and the octahedral [Fe] sublattices. A longitudinal ($\vartheta_m$=0) external magnetic field of $H_{ext}$ = 55 kOe forces the spins to align parallel and antiparallel in the two sublattices. Therefore, the $\Delta m = 0$ lines disappear. Furthermore, the lines corresponding to the more highly populated octahedral sites move inward, indicating a reduction of the effective field, while the weaker lines corresponding to the tetrahedral sites move outward because of an increase of the effective field. This is as expected since the effective field produced at the nucleus by the electronic shells is directed opposite to the atomic moment (negative hyperfine interaction) so that the splitting for the sublattice with the magnetization parallel (dominant octahedral site) or antiparallel (tetrahedral site) to the net moment is decreased or increased, respectively, by the application of an external field.

A surprising effect is produced by a change of just one parameter: the effective thickness, $\tau_A = f_A \sigma_0 Nd$, where $f_A$ is the absorber recoil-free fraction, $\sigma_0$ the resonance cross section, N the number of resonance atoms per $cm^3$ and d the thickness in cm. The spectrum in Fig.6.5 was obtained with an absorber thickness $\tau_A \approx 20$, but the thickness was increased to $\tau_A \approx 360$ for the spectrum of Fig.6.6. Comparison of the spectra in Figs.6.5b,6 suggests that each of the previously well-resolved outer pairs of lines ($\alpha$),[$\alpha$] and ($\eta$),[$\eta$] has been replaced by a three-line pattern. This effect is caused by polarization and overlap. Each line in Fig.6.5b constitutes one specific circularly polarized component as indicated in the stick diagram. Thus, each absorber line by itself can absorb only up to one half of the resonant $\gamma$ rays from an unpolarized source. For the other half, with the opposite helicity, the material is transparent. This phenomenon, where one polarized component (circular or linear) is absorbed while the other component with opposite helicity or linearity is transmitted, is known in optics as dichroism and has practical applications, for instance, in the polaroid sunglasses.

Spectra of $(Mg_{0.26}, Fe_{0.74})[Mg_{0.74}, Fe_{1.26}]O_4$ in a longitudinal ($\vartheta_m$=0°) field of 55 kOe have been computer-simulated for a variety of effective thicknesses as shown in Fig.6.7. It is of interest to investigate the variations in absorption with thickness $\tau_A$ at four Mössbauer velocities especially chosen for their characteristic features. These velocities are designated by the symbols ①, ②, ③, and ④ in Figs. 6.5-7. The resonant absorption at these four velocities as a function of thickness is shown in Fig.6.8.

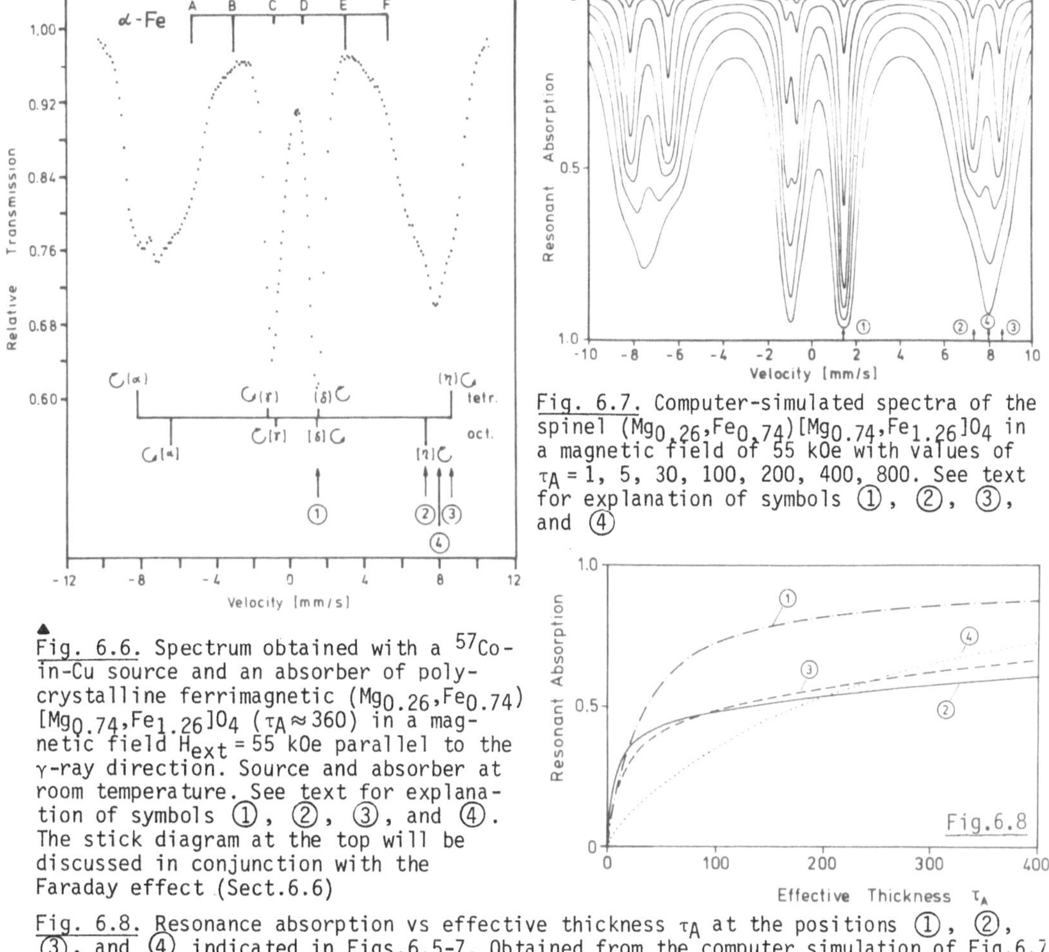


Fig. 6.6. Spectrum obtained with a $^{57}$Co-in-Cu source and an absorber of polycrystalline ferrimagnetic $(Mg_{0.26},Fe_{0.74})[Mg_{0.74},Fe_{1.26}]O_4$ ($\tau_A \approx 360$) in a magnetic field $H_{ext} = 55$ kOe parallel to the γ-ray direction. Source and absorber at room temperature. See text for explanation of symbols ①, ②, ③, and ④. The stick diagram at the top will be discussed in conjunction with the Faraday effect (Sect.6.6)

Fig. 6.7. Computer-simulated spectra of the spinel $(Mg_{0.26},Fe_{0.74})[Mg_{0.74},Fe_{1.26}]O_4$ in a magnetic field of 55 kOe with values of $\tau_A = 1$, 5, 30, 100, 200, 400, 800. See text for explanation of symbols ①, ②, ③, and ④

Fig. 6.8. Resonance absorption vs effective thickness $\tau_A$ at the positions ①, ②, ③, and ④ indicated in Figs.6.5-7. Obtained from the computer simulation of Fig.6.7

At ①, since the positions of the inner lines [δ],(δ) are close together, both polarization components contribute to a large absorption, even though the nuclear transition probabilities are relatively small.

At ②, for a thin sample, the line [η] is the strongest line due to the large transition probabilities for the outer lines and due to the large population of [Fe] in the octahedral site as compared to (Fe) in the tetrahedral site. Because of the dichroism, only 50% of the resonance γ radiation – that is, one circularly polarized component – can be absorbed. Any further increase in resonant absorption at this velocity must come from the overlap of the neighboring (η) line which has the necessary opposite helicity. The population of (Fe) in the tetrahedral sites is small compared to that of [Fe] in octahedral sites and therefore the contribution from overlap by the (η) lines is relatively small. At the opposite extreme, with thick absorbers, the [η] line at ② becomes the weakest line.

At ③, with the outer line (η), we observe the opposite effect. In thin samples the (η) line is weak compared to the [η] line because of lower occupation of the tetrahedral (Fe) site than of the [Fe] octahedral site. With greater absorber thicknesses, however, the strongly occupied octahedral [Fe] site has a considerable resonance overlap effect, absorbing with opposite helicity at the position of the (η) line. Thus, the reversal of the relative intensities of line [η] and line (η) at positions ② and ③ with increasing effective thickness is due to polarization.

At ④, with thick absorbers, a curious effect occurs which gives the impression of an additional resonance line between the two resonance lines [η] and (η) as seen in Figs.6.6,7. In reality, this "line" is only the result of strong overlap at position ④ where both of the circularly polarization components are absorbed.

### 6.2.2 Electric Quadrupole Interaction

The electric field gradient (EFG) does not affect the $I = 1/2$ ground level which has no quadrupole moment and therefore remains twofold degenerate. The $I = 3/2$ excited level is split into two twofold degenerate energy levels. The eigenvalues, eigenstates and transitions are given in Fig.6.3 on the right. Here only one of the two density matrices has to be calculated explicitly since the sum of the two density matrices must be a multiple of the unit matrix (unpolarized isotropic radiation when the two lines are added),

$$\hat{\rho}_{A,\alpha} = \begin{pmatrix} \rho_{-1-1} & \rho_{-11} \\ \rho_{1-1} & \rho_{11} \end{pmatrix}$$

$$\hat{\rho}_{B,\beta} = \frac{1}{2}\hat{E} - \hat{\rho}_{A,\alpha} \quad . \tag{6.36}$$

The lines are partially linearly polarized and do not contain any circular component

$$\rho_{11} = \rho_{-1-1} = \frac{1}{4} + \frac{1}{16}\sqrt{\frac{3}{3+\eta^2}}\,(3\cos^2\vartheta - 1 + \eta\,\sin^2\vartheta\,\cos 2\psi)$$

$$\rho_{1-1} = \rho^*_{-11} = -\frac{1}{16}\sqrt{\frac{3}{3+\eta^2}}\,e^{-i2\varphi}\left\{3\sin^2\vartheta + \eta[(1+\cos^2\vartheta)\cos 2\psi - 2i\,\cos\vartheta\,\sin 2\psi]\right\} \quad . \tag{6.37}$$

When the EFG does not have axial symmetry ($\eta \neq 0$), the intensity and polarization will also depend on the third Euler angle $\psi$ which gives the rotation around the EFG's z-principal axis.

The relative line intensity, as a function of the angle $\vartheta_q$ between the principal axis of the EFG ($\eta=0$) and the $\gamma$-ray direction, can be obtained from Fig.6.1 by combining the two ($\pm 1/2 \to \pm 1/2$) transitions. The ratio $R_q$ of the intensities corresponding to the A(α)($\pm 3/2 \leftrightarrow \pm 1/2$) and B(β)($\pm 1/2 \leftrightarrow \pm 1/2$) lines is given by

$$R_q = \frac{I_{\pm 3/2}}{I_{\pm 1/2}} = \frac{1+\cos^2\vartheta_q}{2/3+\sin\vartheta_q} \quad . \tag{6.38}$$

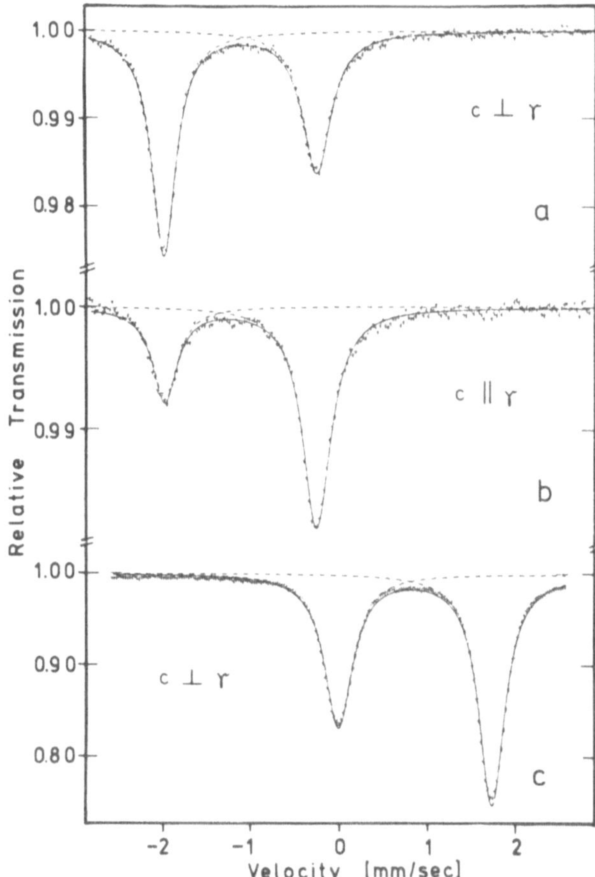


Fig. 6.9a-c. Spectra of single-crystal $LiNbO_3$-$^{57}Co$ source with a $Na_4Fe(CN)_6 \cdot 10\ H_2O$ absorber (a,b) and a $LiNbO_3$-$^{57}Fe$ absorber with a $Pd^{57}Co$ source (c). The orientations of the c-axes of the $LiNbO_3$ crystals with respect to the γ-ray directions are indicated

At $\vartheta_q = 54^\circ 44'$ or for random orientations (disregarding any Goldanskii-Karyagin effect) the two lines have equal intensities.

Co and Fe are quite easily incorporated into $LiNbO_3$, a useful high-resolution optical information-storage material. Annealing in a reducing atmosphere leaves the Co and Fe in the two-valent state [6.16]. The spectra in Fig.6.9 were obtained with a single crystal $LiNbO_3$-$^{57}Co$ source (a,b) or a $LiNbO_3$-$^{57}Fe$ absorber (c) in conjunction with a single line $Na_4Fe(CN)_6 \cdot 10\ H_2O$ absorber or $^{57}CoPd$ source, respectively. The orientations of the c-axis of the $LiNbO_3$ crystals with respect to the γ-ray direction are indicated. The quadrupole coupling constant is negative ($q<0$). The source and the absorber spectra in (a) and (c) are effectively a mirror-image pair. Upon rotating the source by 90° (Fig.6.9a to Fig.6.9b), the two extreme predicted values of $R_q$ (3:5 and 3:1) are nicely verified. This indicates that the principal axis of the EFG coincides with the c-axis and also that the asymmetry parameter $\eta = 0$, or in other words, that the site symmetry of the Co and Fe atoms is axially symmetric.

## 6.3 Polarized $\gamma$ Rays

Three methods are used to produce polarized $\gamma$ radiation from the first excited state of $^{57}$Fe.

### 6.3.1 Magnetized Ferromagnetic Sources

This technique was first demonstrated in 1960 [6.11,12] and it has remained the most common way of producing polarized $\gamma$ rays [6.17]. The emitted radiation pattern of a magnetized source of $^{57}$Co in $\alpha$-Fe consists of six lines, all linearly polarized if observed perpendicular to the direction of $H_{int}$ ($\vartheta_m$=90°). The plane of polarization of the $\gamma$ rays from the transitions corresponding to the $\Delta m = \pm 1$ is perpendicular to the plane of polarization of the $\gamma$ rays from the transitions corresponding to the $\Delta m = 0$. If observed parallel to the direction of $H_{int}$ ($\vartheta_m$=0°), however, the radiation consists of only four lines because the transitions corresponding to $\Delta m = 0$ are missing. The $\gamma$ radiation of these four lines is circularly polarized. The helicity of lines A and D is opposite to that of lines C and F. The helicity of the $\gamma$ radiation reverses with the reversal of the external magnetic field. A $\gamma$ ray is called right circularly polarized if the spin lies in the direction of observation. If $\vartheta_m$ is between 0° and 90°, the observed radiation pattern is elliptically polarized.

### 6.3.2 Quadrupole Split Sources

The usefulness of a quadrupole split $^{57}$Co source depends on whether the source matrix material has the following properties:

1) It can be grown as a single crystal, and the cutting of plates in various directions is possible.
2) $^{57}$Co can be incorporated, with a unique charge state, into a crystallographically unique lattice site.
3) The lattice site has axial symmetry ($\eta$=0) with unique orientation.
4) The quadrupole splitting is large or at least well resolved.

Note that the vibrational isotropy is not required. For an appropriate orientation in a single crystal arrangement *one* Debye-Waller factor is involved and will determine the intensities of both quadrupole split lines. On the other hand, in polycrystalline materials any vibrational anisotropy will cause unusual relative intensities (Goldanskii-Karyagin effect).

Unfortunately, only a very few Fe compounds do fulfill the conditions listed above. If the principal axis of the EFG is perpendicular to the $\gamma$-ray direction the line corresponding to the $\pm 3/2 \leftrightarrow \pm 1/2$ transition is linearly polarized while the line corresponding to the $\pm 1/2 \leftrightarrow \pm 1/2$ transition is partially linearly polarized in the opposite direction. It has been shown that single crystals of Be-$^{57}$Co and

$LiNbO_3$-$^{57}$Co (see the spectra in Fig.6.9) are useful as sources of polarized γ rays [6.18,19]. In the case of Be-$^{57}$Co, the spectrum can be improved or simplified by using a filter to eliminate the unwanted second line.

### 6.3.3 Filter Techniques

To construct a polarizer one can also make use of the phenomenon of dichroism (see Sect.6.1.3) in which certain polarization components are selectively absorbed out of a beam of electromagnetic radiation. For instance, if γ radiation from a single-line non-polarized source is brought into resonance with one of the six absorption lines of a polarizer consisting of a transversely ($\vartheta_m=90^\circ$) magnetized α-Fe foil, or with the absorption line corresponding to the $\pm 3/2 \leftrightarrow \pm 1/2$ transition of a polarizer consisting of a quadrupole-split single crystal which has the principal axis of its EFG perpendicular to the γ-ray direction ($\vartheta_q=0$), the transmitted γ rays will be partially plane polarized [6.17,20,21]. The planes of polarization of the absorbed and transmitted γ radiation are perpendicular to each other. Circularly polarized γ rays may be obtained by using a longitudinally magnetized ferromagnetic or ferrimagnetic material as polarizer in conjunction with a single line source [6.22]. The occurrence of circular dichroism, i.e., the filtering out of polarized γ rays with a certain helicity has been demonstrated in the spectra of the spinel $(Mg_xFe_{1-x})[Mg_{1-x}Fe_{1+x}]O_4$ (Figs.6.6,7) [6.8]. To facilitate the procedure of producing polarized γ rays, either an appropriate isomer shift is selected [6.20,21] or else a double-motion Mössbauer spectrometer is used [6.17] so that the single line source moves with constant velocity in resonance with one of the polarizer lines. The analyzer then moves with variable velocity to cover the whole spectrum.

A variation of the filter technique has proved useful. After selective excitation of one nuclear sublevel in the filter the reemission constitutes a single line with a polarization corresponding to the magnetization [6.23].

## 6.4 Polarimetry

In Mössbauer polarimetry the source and the absorber take the place of the polarizer and analyzer used in optical polarimetry, as shown schematically in Fig.6.10. The degree of absorption in the analyzer depends on the polarization of the γ rays, just as in optical polarimetry. However, in the case of polarized recoil-free γ rays, one has to consider the multitude of individual lines having different polarizations. The locations of the lines in the Mössbauer spectrum are easily calculated by subtracting the line positions of the corresponding hyperfine patterns for the source and absorber. The relative intensities of the lines can be obtained from expressions

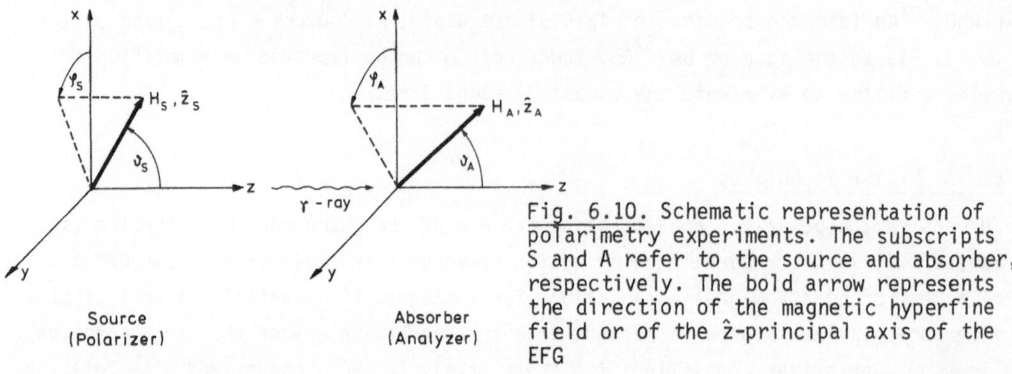


Fig. 6.10. Schematic representation of polarimetry experiments. The subscripts S and A refer to the source and absorber, respectively. The bold arrow represents the direction of the magnetic hyperfine field or of the $\hat{z}$-principal axis of the EFG

derived by FRAUENFELDER at al. [6.13] or from the density matrices given in Sects. 6.3.1,2.

In Fig.6.10 and in what follows, the z-direction has been chosen to be the $\gamma$-ray direction, and the x- and y-directions have then been arbitrarily oriented within the plane perpendicular to z. The subscripts S and A refer to the source and absorber, respectively. For both the source (polarizer) and the absorber (analyzer), the direction of the magnetic field H or the principal axis of the quadrupole EFG $\hat{z}$ is represented by the bold arrow. $\vartheta$ is the polar angle between H or $\hat{z}$ and the $\gamma$-ray direction (z-direction). $\varphi$ is the azimuthal angle between the x-direction and the projection of H or $\hat{z}$ on the x-y plane.

For simplicity we will discuss some cases involving sources (polarizers) which emit circularly or linearly polarized $\gamma$ rays, i.e., $H_S$ or $\hat{z}_S$ parallel, antiparallel or perpendicular to the $\gamma$ ray direction. The relative intensities of the various lines of the resulting Mössbauer spectra for a variety of absorber (analyzer) polarizations are shown in Table 6.2.

### 6.4.1 Circularly Polarized $\gamma$ Rays

First let us consider the case of circularly polarized $\gamma$ rays emitted and absorbed in ferromagnetic materials [6.24]. Conservation of angular momentum requires that for absorption, the helicity of the radiation from the source has to be matched in the absorber. If source and absorber are placed in collinear longitudinal magnetic fields ($\vartheta_S$=0,$\vartheta_A$=0), eight lines are expected for both parallel and antiparallel fields. The positions and the intensities of the lines for a thin absorber as a function of the ratio of the hyperfine fields in the source and absorber, $H_A/H_S$, are represented in the nomograph at the bottom of Fig.6.11 where a hyperfine field of $H_S \approx 300$ kOe has been assumed. The four Mössbauer lines at the center, $H_A/H_S = 0$, correspond to the condition that the absorber absorbs a single line. If $H_S = H_A$ ($H_A/H_S$=1), the fields are parallel to each other, and because of coincidences of source and absorber line positions, only three Mössbauer lines occur. For equal but

Table 6.2. Relative line intensities for Mössbauer spectra. Obtained with a source of linearly or circularly polarized $\gamma$ rays and a thin absorber which is either magnetically ordered along a unique direction or is a single crystal with an axially symmetric ($\eta$=0) quadrupole EFG whose principal axis $\hat{z}$ is along a unique direction. Refer to Fig.6.10. For more general orientations or EFG with $\eta \neq 0$ make use of the density matrices given in (6.35,37)

**Circularly polarized $\gamma$ rays**

$H_S \to H_A$ Magnetized source, magnetized absorber

| $\vartheta_{S,A} = 0^\circ$, $\Delta\varphi$ indeterminate | $\alpha$ | $\gamma$ | $\delta$ | $\eta$ | $\vartheta_S = 0^\circ$, $\vartheta_A = 180^\circ$, $\Delta\varphi$ indeterminate | $\alpha$ | $\gamma$ | $\delta$ | $\eta$ |
|---|---|---|---|---|---|---|---|---|---|
| A | 9 | 0 | 3 | 0 | A | 0 | 3 | 0 | 9 |
| C | 0 | 1 | 0 | 3 | C | 3 | 0 | 1 | 0 |
| D | 3 | 0 | 1 | 0 | D | 0 | 1 | 0 | 3 |
| F | 0 | 3 | 0 | 9 | F | 9 | 0 | 3 | 0 |

**Linearly polarized $\gamma$ rays**

a) $H_S \to H_A$ Magnetized source, magnetized absorber

$\vartheta_S = 90^\circ$, $\vartheta_A$ and $\Delta\varphi$ variable

| | $\alpha,\eta$ | $\beta,\varepsilon$ | $\gamma,\delta$ |
|---|---|---|---|
| A,F | $9(1-\sin^2\vartheta_A \sin^2\Delta\varphi)$ | $12 \sin^2\vartheta_A \sin^2\Delta\varphi$ | $3(1-\sin^2\vartheta_A \sin^2\Delta\varphi)$ |
| B,E | $12(1-\sin^2\vartheta_A \cos^2\Delta\varphi)$ | $16 \sin^2\vartheta_A \cos^2\Delta\varphi$ | $4(1-\sin^2\vartheta_A \cos^2\Delta\varphi)$ |
| C,D | $3(1-\sin^2\vartheta_A \sin^2\Delta\varphi)$ | $4 \sin^2\vartheta_A \sin^2\Delta\varphi$ | $(1-\sin^2\vartheta_A \sin^2\Delta\varphi)$ |

| | $\vartheta_{S,A} = 90^\circ$, $\Delta\varphi = 0^\circ$ | | | $\vartheta_{S,A} = 90^\circ$, $\Delta\varphi = 90^\circ$ | | | $\vartheta_S = 90^\circ$, $\vartheta_A = 0^\circ$, $\Delta\varphi$ indeterminate | | |
|---|---|---|---|---|---|---|---|---|---|
| | $\alpha,\eta$ | $\beta,\varepsilon$ | $\gamma,\delta$ | $\alpha,\eta$ | $\beta,\varepsilon$ | $\gamma,\delta$ | $\alpha,\eta$ | $\beta,\varepsilon$ | $\gamma,\delta$ |
| A,F | 9 | 0 | 3 | 0 | 12 | 0 | 9 | 0 | 3 |
| B,E | 0 | 16 | 0 | 12 | 0 | 4 | 12 | 0 | 4 |
| C,D | 3 | 0 | 1 | 0 | 4 | 0 | 3 | 0 | 1 |

b) $H_S \to \hat{z}_A$ Magnetized source, single-crystal absorber

| | $\vartheta_{S,A} = 90^\circ$, $\Delta\varphi = 0^\circ$ | | $\vartheta_{S,A} = 90^\circ$, $\Delta\varphi = 0^\circ$ | | $\vartheta_{S,A} = 0^\circ$, $\Delta\varphi$ indeterminate | |
|---|---|---|---|---|---|---|
| | $\alpha$ | $\beta$ | $\alpha$ | $\beta$ | $\alpha$ | $\beta$ |
| A,F | 9 | 3 | 0 | 12 | 9 | 3 |
| B,E | 0 | 16 | 12 | 4 | 12 | 4 |
| C,D | 3 | 1 | 0 | 4 | 3 | 1 |

c) $\hat{z}_S \to \hat{z}_A$ Single crystal source, single-crystal absorber

| | $\vartheta_{S,A} = 90^\circ$, $\Delta\varphi = 0^\circ$ | | $\vartheta_{S,A} = 90^\circ$, $\Delta\varphi = 90^\circ$ | | $\vartheta_{S,A} = 0^\circ$, $\Delta\varphi$ indeterminate | |
|---|---|---|---|---|---|---|
| | $\alpha$ | $\beta$ | $\alpha$ | $\beta$ | $\alpha$ | $\beta$ |
| A | 9 | 3 | 0 | 12 | 9 | 3 |
| B | 3 | 17 | 12 | 8 | 15 | 5 |

opposite fields ($H_A/H_S$=-1), six lines are expected. The spectrum in Fig.6.11 was taken with a source of $^{57}$Co in $\alpha$-Fe and a $\alpha$-Fe absorber both in an external magnetic field of 52 kOe applied perpendicular to the planes of the foils. The external applied field in conjunction with the demagnetizing field and the negative hyperfine interaction reduces the internal field to about 300 kOe.

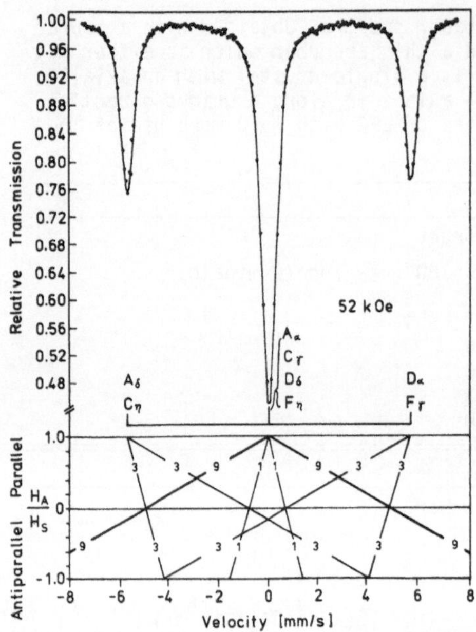


Fig. 6.11. Spectrum obtained with a source of $^{57}$Co in α-Fe and an α-Fe absorber. Both source and absorber are in a longitudinal magnetic field ($\vartheta_S=\vartheta_A=0^\circ$) $H_{ext}=52$ kOe. The stick diagram, the letters, and the nomogram at the bottom indicate the positions, relative line intensities and origins of the lines. $H_A/H_S=1$ represents the experimental situation for the spectrum shown

### 6.4.2 Linearly Polarized γ Rays (Magnetic Hyperfine Interaction)

For a source of $^{57}$Co in α-Fe which is magnetized perpendicular to the γ-ray direction ($\vartheta_S=90^\circ$), all six hyperfine lines are linearly polarized. A magnetically ordered absorber will show, in general, 36 absorption lines when used with such a linearly polarized source. The differences of the positions of the source and absorber lines give the locations of the lines of the Mössbauer spectrum. If the spins in both the source and absorber are oriented perpendicular to the γ-ray direction ($\vartheta_S=\vartheta_A=90^\circ$), two limiting cases of parallel ($H_S||H_A$) and perpendicular ($H_S\perp H_A$) magnetic fields with Mössbauer patterns of 20 or 16 lines, respectively, can be distinguished. Assuming a hyperfine field of $H_{int}=330$ kOe for the source, the positions and the intensities of the lines of the Mössbauer spectrum as a function of the ratio $H_A/H_S$ of the hyperfine fields in absorber and source are as shown in the nomograph of Fig.6.12. In the center, at $H_A/H_S=0$, the positions correspond to the anticipated six-line pattern for an α-Fe source and an unsplit absorber. In the cases $H_S=\pm H_A$, coincidences of source and absorber lines occur and nine or six lines, respectively, are expected. The spectra in Fig.6.12 were obtained with a source of $^{57}$Co in α-Fe and an absorber of α-Fe, both at room temperature and magnetized perpendicular to the γ-ray direction. The stick diagram indicates the positions and relative line intensities for (a) $H_S||H_A$, (b) $H_S\perp H_A$. The Roman and Greek letters designate the origin of the lines (see Fig.6.3).

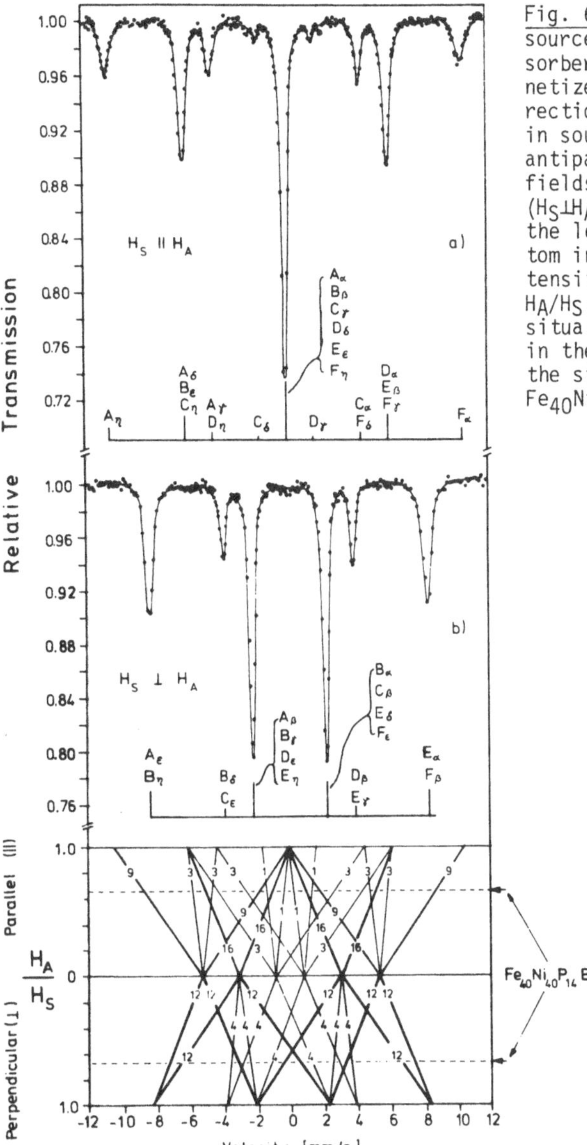


Fig. 6.12a,b. Spectra obtained with a source of $^{57}$Co in $\alpha$-Fe and an $\alpha$-Fe absorber both at room temperature and magnetized perpendicular to the $\gamma$-ray direction ($\vartheta_S=\vartheta_A=90^\circ$). (a) Magnetic fields in source and absorber are parallel or antiparallel ($H_S \| H_A; \varphi_S=\varphi_A$). (b) Magnetic fields are perpendicular to each other ($H_S \perp H_A$; $\varphi_S-\varphi_A=90^\circ$). The stick diagram, the letters and the nomogram at the bottom indicate the positions, relative intensities and origins of the lines. $H_A/H_S = 1$ represents the experimental situation for both spectra. The arrows in the nomogram at $\bar{H}_A/H_S \approx 0.63$ indicate the situation for the amorphous metal $Fe_{40}Ni_{40}P_{14}B_6$ (Figs.6.14,15)

### 6.4.3 Linearly Polarized $\gamma$ Rays (Quadrupole Interaction)

In general, if both source and absorber have quadrupole-split spectra one can expect four Mössbauer lines. Under the condition that the magnitude of the quadrupole splitting is the same, the number of lines reduces to three because of coincidences of source and absorber line patterns. Spectra of a $LiNbO_3$-$^{57}$Co source vs a $LiNbO_3$-$^{57}$Fe absorber with their crystallographic c-axes perpendicular to the $\gamma$-ray direction ($\vartheta_S=\vartheta_A=90^\circ$) are presented in Fig.6.13(a) at the left. In this material the principal axis of the EFG coincides with the c-axis [6.16]. For the upper spectrum, the c-axes

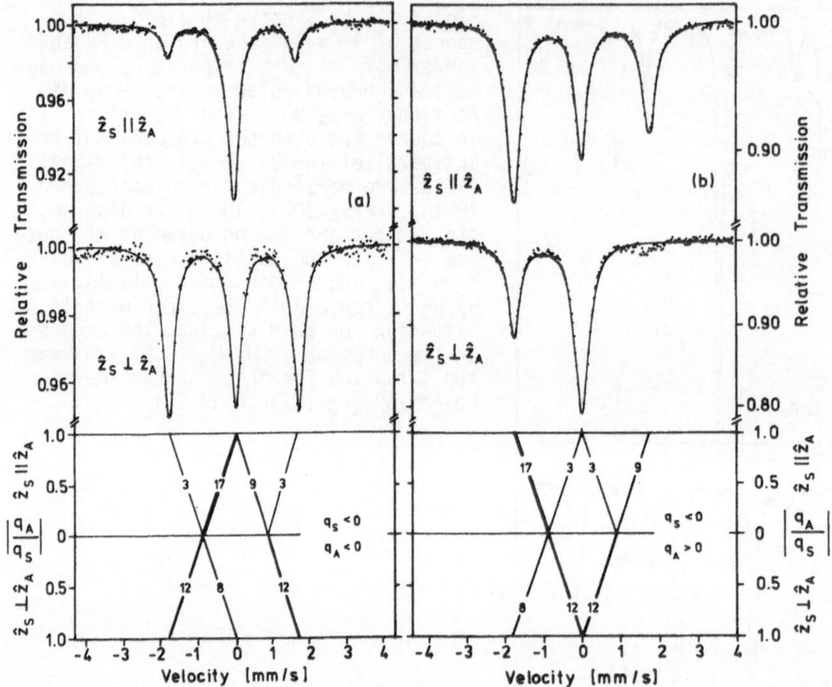

Fig. 6.13a,b. Spectra obtained with a single-crystal $LiNbO_3$-$^{57}Co$ source oriented with the c-axis perpendicular to the $\gamma$-ray direction ($\vartheta_S=90^o$) and single-crystal absorbers of $LiNbO_3$-$^{57}Fe$ (a) and $FeCO_3$ (b) with c-axis orientation $\vartheta_A = 90^o$. For the upper spectra the c-axes were parallel ($\hat{z}_S \| \hat{z}_A$) and in the lower they were perpendicular ($\hat{z}_S \perp \hat{z}_A$) to each other. The nomograms at the bottom indicate the positions and relative line intensities. $q_A/q_S \approx 1$ represents the experimental situation for the spectra shown

were parallel ($\hat{z}_S \| \hat{z}_A$), and for the lower spectrum, the c-axes were perpendicular ($\hat{z}_S \perp \hat{z}_A$) to each other.

At the bottom of Fig.6.13, nomograms indicate the positions and intensities of the Mössbauer lines expected for a thin absorber with parallel ($\varphi_A=\varphi_S$) and perpendicular ($\varphi_A-\varphi_S=90^o$) arrangements of the EFG principal axes in the source and absorber as a function of the absolute value $q_A/q_S$ of the ratio of the EFG's. The velocity is scaled for $LiNbO_3$:$^{57}Co$, $^{57}Fe$ ($\Delta E_Q$=1.74 mm/s).

For a polarimeter with $LiNbO_3$-$^{57}Co$ as source (polarizer), the mineral siderite ($FeCO_3$) is of particular interest as an absorber (analyzer) [6.19]. Fe in $FeCO_3$ (rhombohedral structure) has axial symmetry, and natural single crystals of this mineral are available. Furthermore, $FeCO_3$ exhibits at room temperature a quadrupole splitting $\Delta E_Q = 1.798 \pm 0.004$ mm/s which is similar to the value for $LiNbO_3$-$^{57}Fe$ ($\Delta E_Q$=1.74±0.02 mm/s). However, in contrast to $LiNbO_3$-$^{57}Fe$, the quadrupole coupling constant is positive ($q_A>0$). Spectra for single crystals of $LiNbO_3$-$^{57}Co$ vs $FeCO_3$ are shown in Fig.6.13(b), again with c-axes oriented perpendicular to the $\gamma$-ray direction ($\vartheta_A=\vartheta_S=90^o$). The two axes were parallel ($\varphi_A=\varphi_S$) in the upper spectrum, and they were perpendicular to each other ($\varphi_A-\varphi_S=90^\circ$) in the lower spectrum. Positions and intensities are again indicated in the nomogram at the bottom. Of particular interest is the line on the positive velocity-side at about 1.8 mm/s because it results from the $\pm3/2 \leftrightarrow \pm1/2$ transitions in both the source and the absorber and involves linear polarization in both emission and absorption. In the perpendicular arrangement, the $\gamma$ rays from this transition in the source have the wrong linear polarization for absorption by the corresponding transition in the absorber which is therefore transparent to these $\gamma$ rays. Only a small residual absorption contribution can be detected which is most likely due to imperfections in the mineral $FeCO_3$ single crystal or to a misalignment of the polarimeter.

### 6.4.4 Special Applications (Amorphous Metals)

For a source (polarizer) which emits $\gamma$ rays that are circularly or linearly polarized by magnetic or quadrupole hyperfine interactions, Table 6.2 is helpful in determining the orientation of the spins or of the principal axis of the EFG in the absorber (analyzer). The method consists of an analysis of the relative intensities of the Mössbauer lines obtained from a single crystal cut in an arbitrary direction and rotated through various angles with respect to the polarizer. This method was first applied, in the case of a quadrupole interaction, to $FeSiF_6 \cdot 6H_2O$ [6.25] and, in the case of a magnetic interaction to Ca(Fe)[Fe]$O_5$ [6.26]. Since then the technique has become more sophisticated [6.27-30] and certain difficulties, particularly in the determination of the local EFG parameters, have also been pointed out [6.31-33]. The materials, properties, and phenomena which have been investigated are: $Fe_3(PO_4)_2 \cdot 8H_2O$ (Vivianite) [6.34], $Ca_2FeAlO_5$ [6.35,36], $FeCO_3$ (Siderite) [6.37,38], $\alpha$-FeOOH (Goethite) [6.39], $\gamma$-FeOOH (Lepidocrocite) [6.40], $ErFeO_3$ (spin reorientation) [6.41], FeOCl [6.42,43], $NH_4Fe(SO_4)_2 \cdot 12H_2O$ (anisotropic relaxation) [6.44], deoxy-myoglobin [6.33], $(NH_4)_2Fe(SO_4)_2 \cdot 6H_2O$ [6.45], $Na_2Fe(CN)_5(NO) \cdot 2H_2O$ [6.46],

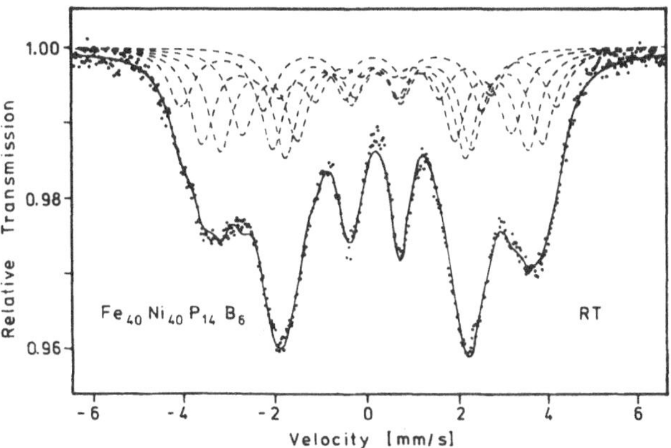


Fig. 6.14. Spectrum of a ribbon of the amorphous metal $Fe_{40}Ni_{40}P_{14}B_6$ at room temperature. Concerning the fittings, see text

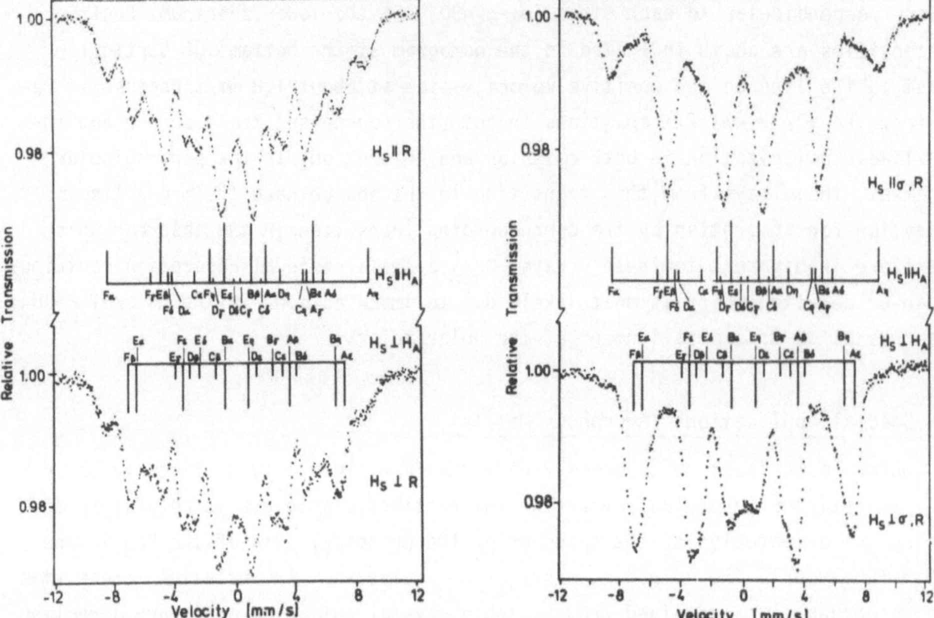


Fig. 6.15. Spectra obtained with a transversely magnetized source of $^{57}Co$ in $\alpha$-Fe ($\vartheta_S=90^\circ$) and an $Fe_{40}Ni_{40}P_{14}B_6$ absorber with the $\gamma$-ray direction perpendicular to the plane of the ribbon. In the upper spectra the source magnetic field $H_S$ was parallel to the ribbon direction R ($H_S \| R$) and in the lower spectra it was perpendicular ($H_S \perp R$). The spectra on the right side were obtained by applying a tensile stress $\sigma$ along R. The stick diagrams and letters in the center indicate the positions and relative intensities of the broad lines neglecting the subspectrum analysis of Fig.6.14. These resonance conditions are also represented by the arrows at $H_A/H_S \approx 0.63$ in the nomogram of Fig.6.12

texture [6.47,48], recording tapes [6.49], $K_3Fe(CN)_6$ [6.50]. A special example concerning amorphous metals will be discussed in more detail.

Ferromagnetic amorphous metals of the type $T_{80}M_{20}$ (T=transition metal, M=metalloid) have attracted considerable interest in recent years, particularly due to the potential applications of these materials. The spectrum of amorphous $Fe_{40}Ni_{40}P_{14}B_6$ is shown in Fig.6.14. It exhibits typical broad lines indicating a hyperfine magnetic field distribution with an average field of about $\bar{H}_A \approx 208$ kOe. The spectrum was fitted [6.51] by a superposition of five subspectra representing five different close-contacts of Bernal's geometrical model of a liquid [6.52]. The probabilities of occurrence of these coordinations as predicted by Bernal's model seem to be reflected in the relative line intensities. Spectra taken using linearly polarized $\gamma$ rays from a transversely magnetized ($\vartheta_S=90^\circ$) source of $^{57}Co$ in $\alpha$-Fe are shown in Fig.6.15 [6.53]. The spectra at the top were obtained by an arrangement where the source magnetic field $H_S$ and the orientation of the amorphous ribbon R were parallel ($H_S \| R$). For the spectra at the bottom they were perpendicular to each other ($H_S \perp R$). The stick diagram and the letters in the center parts indicate expected positions and relative line intensities for the Mössbauer spectrum assuming, for simplicity, only six absorption lines corresponding to a certain average hyperfine field $\bar{H}_A$ for the amorphous material so that the subspectra could be ignored. The value chosen for $\bar{H}_A$ was such that $\bar{H}_A/H_S \approx 0.63$, which is marked by two arrows in the nomogram of Fig.6.12. In comparing the spectra on the left in Fig.6.15 with the stick diagrams in the center corresponding to the two polarizations, it becomes evident that both the 20-line ($H_S \| H_A$) and the 16-line ($H_S \perp H_A$) patterns are present in both spectra, although in different relative amounts. Upon inspection it is easy to see that the upper (lower) spectrum $H_S \| R$, ($H_S \perp R$) agrees better with the stick diagram for $H_S \| H_A$, ($H_S \perp H_A$). Therefore, one can conclude that the spins in the amorphous metal are preferentially orientated in the ribbon direction R. In a detailed analysis of this problem, one can calculate a so-called minimum texture, in the sense that the true distribution of spin orientations may have additional structure, but this method, based on the angular dependence of the hyperfine interaction, is inherently incapable of detecting it [6.48,54].

The spectra on the right-hand side in Fig.6.15 were obtained by applying a tensile stress along the ribbon direction R. Clearly, this stress has caused a large effect. Now, by comparing the spectra obtained in the experimental arrangements $H_S \| \sigma$, R and $H_S \perp \sigma$, R with the stick diagrams in the center, it is seen that the upper (lower) spectrum matches the corresponding orientation $H_S \| H_A$ ($H_S \perp H_A$) very well and also the choice of $\bar{H}_A/H_S \approx 0.63$. Because the spectral lines above and below have become mutually exclusive, the conclusion can be drawn that the stress $\sigma$ has fully polarized the absorber lines and, therefore, the spins have been aligned into the direction of the stress. A good fit to the spectra has been obtained by taking into account 6 (polarized source lines, Roman capital letters) $\times$ 6 (broad absorber lines, Greek letters) $\times$ 5 (absorber subspectra according to the Bernal model) = 180 lines in the Mössbauer spectrum. With polarization, the upper spectrum contains 100 lines and the lower spectrum 80 lines.

## 6.5 $\gamma$-Ray Rotation Polarimeter

A Mössbauer polarimeter consisting of a polarizer (source) of $^{57}$Co in $\alpha$-Fe and an $\alpha$-Fe analyzer (absorber), both polarized transversely to the $\gamma$-ray direction by small magnets ($\vartheta_S = \vartheta_A = 90^\circ$), is shown schematically in Fig.6.16 [6.55] (see also [6.17,20,21,37,56]). There is no relative Doppler motion. The counting rate R in the detector depends only on the angle $\Delta\varphi = \varphi_S - \varphi_A$ between the two applied magnetic fields. The resulting sinusoidal behavior, as shown in Fig.6.17(a), is called in optics the Malus curve:

$$R(\Delta\varphi) = R_0 - R_1[1+\cos(2\Delta\varphi+2\phi)] \quad . \tag{6.39}$$

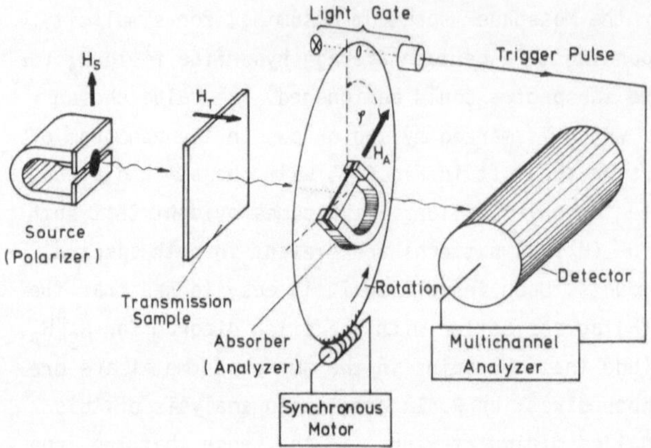


Fig. 6.16. Automatic Mössbauer polarimeter arrangement

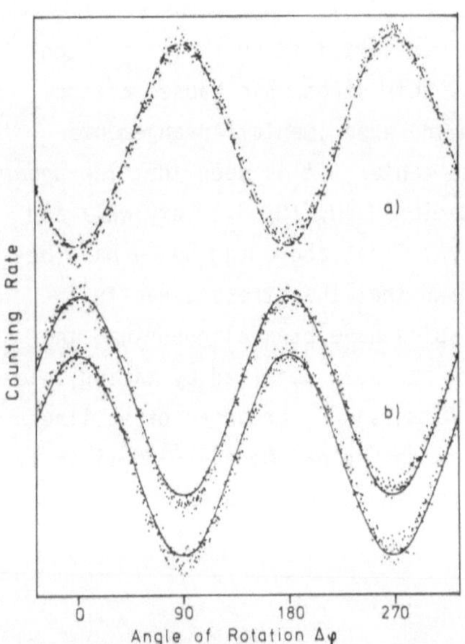


Fig. 6.17a,b. Counting rate vs rotation angle $\Delta\varphi$ between polarizer and analyzer: Malus curve. (a) Stationary source (v=0); (b) moving source (v=± 8.41 mm/s) and "optically active" transmission sample

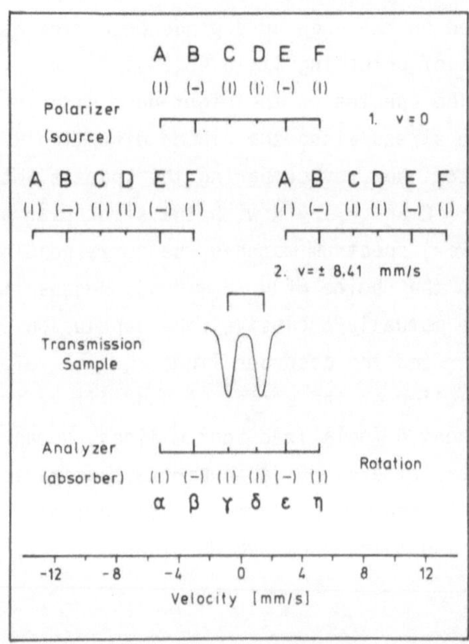


Fig. 6.18. Positions, polarization and motion of the source (polarizer) and absorber (analyzer) in a polarimeter. The Mössbauer spectrum of an optically active transmission sample enriched in $^{57}$Fe is shown

$R_0$ denotes the background and nonresonant component and $R_1$ is the amplitude of the Malus curve. Birefringence rotation $\phi$ might be caused by transmission through a sample placed between the polarizer and the analyzer. In the automatic polarimeter of Fig.6.16, the absorber, with its permanent magnet, rotates continuously, driven by a synchronous motor. The rotation frequency coincides with the sweep frequency of a multichannel analyzer which is triggered by a light pulse gated by the rotating absorber holder. Thus, each channel of the multichannel analyzer corresponds to a certain angular position $\Delta\varphi$ of the source relative to the absorber. In operating the spectrometer, the "optical axis" has to be well adjusted to avoid geometrical effects. The upper curve (a) in Fig.6.17 was obtained by rotating the $\alpha$-Fe absorber of the polarimeter, while the source of $^{57}$Co in $\alpha$-Fe was stationary. For the lower curves (b) the source was moved at two constant Doppler velocities, $v = \pm 8.41$ mm/s. (In this case a transmission sample was placed between the polarizer and analyzer, but as will be seen, this had little or no effect on the result, and so its presence can be ignored for the time being.) The line positions, polarizations, and the Doppler motion of the source (polarizer) are indicated at the top of Fig.6.18. The line positions and polarizations of the absorber (analyzer) are indicated at the bottom of Fig.6.18. (The absorption pattern for the transmission sample, at the center of Fig. 6.18, may be ignored for now.) At zero Doppler velocity, the minimum and the maximum of the Malus curve correspond to the two conditions at zero velocity in Fig.6.12 where strong absorption ($H_S \| H_A, \Delta\varphi=0^\circ$) and total transmission ($H_S \perp H_A, \Delta\varphi=90^\circ$), respectively, are observed.

If the source is moved at one of two constant Doppler velocities, $v = \pm 8.41$ mm/s, the source lines A and B or E and F are brought into resonance with the absorber lines $\varepsilon$ and $\eta$ or $\alpha$ and $\beta$, respectively (see Fig.6.18). However, in this case maximum resonance absorption of the linearly polarized source lines by the linearly polarized absorber lines (A$\varepsilon$ and B$\eta$ or E$\alpha$ and F$\beta$) occurs if $H_S$ and $H_A$ are oriented perpendicular to each other ($H_S \perp H_A; \Delta\varphi=90^\circ$). These $90^\circ$ shifts can be seen in the two lower Malus curves of Fig.6.17.

## 6.6 Birefringence Polarimetry

The dispersion associated with Mössbauer resonance absorption in a transmission sample can be measured by birefringence rotation polarimetry [6.17,57-65]. Here we want to discuss experiments on the Mössbauer Faraday effect and "optical" rotation.

### 6.6.1 Mössbauer Faraday Effect

In Sect.6.1.3 a formal description was given of the changes in polarization and intensity to be expected when a $\gamma$ ray passes through a polarized medium. One particular feature of these phenomena, which in optics is known as Faraday rotation, will be illustrated by a practical demonstration. Consider the experimental setup shown schematically in Fig.6.16 consisting of a $\gamma$-ray source (polarizer) of $^{57}$Co in $\alpha$-Fe, a transmission sample, an $\alpha$-Fe (analyzer), and a detector. The source is magnetized perpendicular to the $\gamma$-ray direction and produces linearly polarized $\gamma$ rays. If the transmission sample were absent, the $\gamma$ ray would reach the absorber directly, part of it being absorbed and the rest being counted in the detector. As discussed above, the intensity absorbed depends on the direction of magnetization of the absorber and thus may be rotated about the "optical" axis.

In the following, the intensity reaching the detector will be derived using the formalism of Sect.6.1.3 and taking into account only the resonant part of the radiation.

We first consider the case where no transmission sample is present. Since the magnetic field in the source is perpendicular to the γ-ray direction, we have $\vartheta_S = 90^\circ$. Substitution in (6.34) gives for the density matrices of the linearly polarized lines B and E,

$$\hat{\rho}_B = \hat{\rho}_E = \frac{1}{8}\begin{pmatrix} 1 & e^{2i\varphi_S} \\ e^{-2i\varphi_S} & 1 \end{pmatrix} \tag{6.40a}$$

and for the density matrices of the emission lines A, C, D and F, which are also linearly polarized, but perpendicular to B and E,

$$\hat{\rho}_A = 3\hat{\rho}_C = 3\hat{\rho}_D = \hat{\rho}_F = \frac{3}{32}\begin{pmatrix} 1 & -e^{2i\varphi_S} \\ -e^{-2i\varphi_S} & 1 \end{pmatrix} \quad . \tag{6.40b}$$

Analogous expressions can be obtained for the absorption density matrices. In the thin-absorber approximation, as has been shown in Sect.6.1.3, the absorbed intensity is proportional to the trace of $\hat{\rho}_0 \cdot \hat{\rho}_{abs}$. In the present case, where the lines are well separated in energy, evaluation of this trace shows that the absorbed intensity is proportional to $1 + \cos(2\Delta\varphi)$ where $\Delta\varphi = \varphi_A - \varphi_S$ is the angle about which the analyzer magnetization has been rotated with respect to that of the polarizer.

This fact may now be exploited to analyze the changes imposed on γ radiation by its passage through a transmission sample located between the polarizer and analyzer. Here, one has to be aware that the change of polarization depends on energy and is thus different for different lines. Also, (which is less obvious) the changes in the γ radiation are not limited to just attenuation and rotation of the plane of polarization. In addition, completely linear polarization may be scattered into elliptical polarization. For a single linearly polarized source line this second effect does not have an influence on the amount of the measured Faraday rotation and so this simpler case will be dealt with first.

The transmission sample is placed in a magnetic field $H_T$ which is parallel to the γ-ray direction so that only four of the six possible transitions are detected. The four lines all have circular polarization with density matrices given by (6.34), with $\vartheta_T = 0^\circ$, as

$$\hat{\rho}_\alpha = 3\hat{\rho}_\delta \propto \begin{pmatrix} 1 & 0 \\ 0 & 0 \end{pmatrix} \tag{6.41}$$

and

$$\hat{\rho}_\eta = 3\hat{\rho}_\gamma = \hat{\rho}_\alpha^x \propto \begin{pmatrix} 0 & 0 \\ 0 & 1 \end{pmatrix} . \tag{6.42}$$

Normalizing these density matrices according to

$$\sum_{\ell=\alpha,\gamma,\delta,\eta} \hat{\rho}_\ell = \hat{E} \tag{6.43}$$

yields for the index of refraction [see (6.24)]

$$\hat{n}kd = kd\hat{E} - \frac{1}{2}\tau \sum_{\ell=\alpha,\gamma,\delta,\eta} \hat{\rho}_\ell \frac{\Gamma/2}{E-E_\ell+i(\Gamma/2)} . \tag{6.44}$$

With this index of refraction, the Poincaré representation of radiation of energy E which was originally linear polarized, but has passed through the transmission samples, is [see (6.26,27)]

$$I = I_0 e^{i2\mathrm{Im}\{a\}} \cosh(2\mathrm{Im}\{b_\zeta\}) \tag{6.45}$$

$$P_\xi = \frac{\cos(2\mathrm{Re}\{b_\zeta\})}{\cosh(2\mathrm{Im}\{b_\zeta\})} \tag{6.46}$$

$$P_\eta = -\frac{\sin(2\mathrm{Re}\{b_\zeta\})}{\cosh(2\mathrm{Im}\{b_\zeta\})} \tag{6.47}$$

$$P_\zeta = -\tanh(2\mathrm{Im}\{b_\zeta\}) , \tag{6.48}$$

where Re and Im denote the real and imaginary part of a complex number.

If one is only interested in the angle of rotation and not, for example, in the degree of polarization, one may proceed as follows. The angle of rotation $\phi$ is defined as that angle $\Delta\varphi = \varphi_A - \varphi_S$ between the magnetic field directions at the polarizer and analyzer for which the count rate in the detector is a minimum. Assuming monochromatic source radiation, $\phi$ is determined by

$$\frac{\partial I(\varphi)}{\partial \Delta\varphi} = \frac{\partial}{\partial \Delta\varphi}\bigg|_\phi \mathrm{Tr}\left\{e^{i\hat{n}kd}\, \hat{\rho}_0\, e^{-i\hat{n}^+kd}\, \hat{\rho}_{\mathrm{analyzer}}\right\} = 0 . \tag{6.49}$$

Using the thin-absorber approximation for the analyzer, one obtains

$$\tan 2\phi = \frac{P_\eta}{P_\xi} \tag{6.50}$$

and the solution

$$\phi = -\mathrm{Re}\{b_\zeta\} = -\frac{\tau_T}{16}\frac{\Gamma}{2}\left[\frac{3(E-E_\alpha)}{(E-E_\alpha)^2+(\Gamma/2)^2} - \frac{(E-E_\gamma)}{(E-E_\gamma)^2+(\Gamma/2)^2} + \frac{(E-E_\delta)}{(E-E_\delta)^2+(\Gamma/2)^2} - \frac{3(E-E_\eta)}{(E-E_\eta)^2+(\Gamma/2)^2}\right] , \tag{6.51}$$

where $\tau_T$ is the total effective thickness of the transmission sample, $\Gamma$ is the full linewidth of the sample, and E is the energy of the monochromatic source radiation.

Taking into account all the six lines in both polarizer and analyzer, the total count rate in the detector as a function of the angle between source and analyzer orientation becomes

$$I(\Delta\varphi) = \text{const} - \sum_{\ell} I_{S\ell} \exp[-2\mathrm{Im}\{a_\ell\}] I_{A\ell} \cos 2(\mathrm{Re}\{b_{\zeta\ell}\}+\Delta\varphi) \quad , \tag{6.52}$$

where $I_{S\ell}$ and $I_{A\ell}$ are the relative line intensities in source and absorber which in our case are proportional to 3:4:1:1:4:3. The expansion coefficients $a_\ell$ and $b_\ell$ of the refractive index of the transmission sample are defined in connection with (6.26) and are to be evaluated at the energy of line $\ell$, that is, $a_\ell = a(E_\ell)$ and $b_{\zeta\ell} = b_\zeta(E_\ell)$, where $E_\ell$ is the energy of line $\ell$.

The total observed Faraday rotation $\phi$ is then given by

$$\tan 2\phi = \frac{\sum_\ell C_\ell \sin 2\phi_\ell}{\sum_\ell C_\ell \cos 2\phi_\ell} \quad , \tag{6.53}$$

where

$$C_\ell = I_{S\ell} I_{A\ell} \exp[-2\mathrm{Im}\{a_\ell\}] \quad \text{and} \quad \phi_\ell = -\mathrm{Re}\{b_{\zeta\ell}\} \quad .$$

It can be seen that when the transmitter has several lattice sites, the resultant rotation obtained from all source and absorber lines and over all lattice sites in the transmitter will, in the general case, not be just the sum of all rotations caused by all the individual lattice sites.

Although Mössbauer Faraday experiments and other $\gamma$-ray electro- and magneto-optics applications were suggested soon after the discovery of the Mössbauer effect [6.66], very few experiments of this type have actually been carried out up to now. The materials studied so far are $^{57}Fe$ in ferromagnetic $\alpha$-Fe [6.17], ferrimagnetic $MgFe_2O_4$ [6.8], paramagnetic $Fe^{2+}$ in MgO [6.57], and FeV [6.62]. $^{99}Ru$ in $\alpha$-Fe has also been investigated [6.10].

The Mössbauer experiments on highly enriched inverse spinel $(Mg_{0.26}Fe_{0.74})[Mg_{0.74}Fe_{1.26}]O_4$ are instructive. Spectra of this spinel with large effective thickness ($\tau_A \approx 360$) are shown in Figs.6.6,7. This material was placed in a longitudinal ($\vartheta_T = 0^\circ$) magnetic field of $H_T = 55$ kOe between a source of $^{57}Co$ in $\alpha$-Fe (polarizer) and an $\alpha$-Fe absorber (analyzer), both transversely magnetized ($\vartheta_S = \vartheta_A = 90^\circ$) as shown schematically in Fig.6.16.

Figure 6.19 shows the Faraday rotatory power of $(Mg_{0.26};Fe_{0.74})[Mg_{0.74};Fe_{1.26}]O_4$ with effective thickness $\tau_A \approx 360$ in a magnetic field of $H_T = 55$ kOe as computed for a single linearly polarized source line of natural width, neglecting absorption. Close to resonance energies, where the dispersion changes sign, the rotatory power may become very large. However, for these energies absorption is also large, with

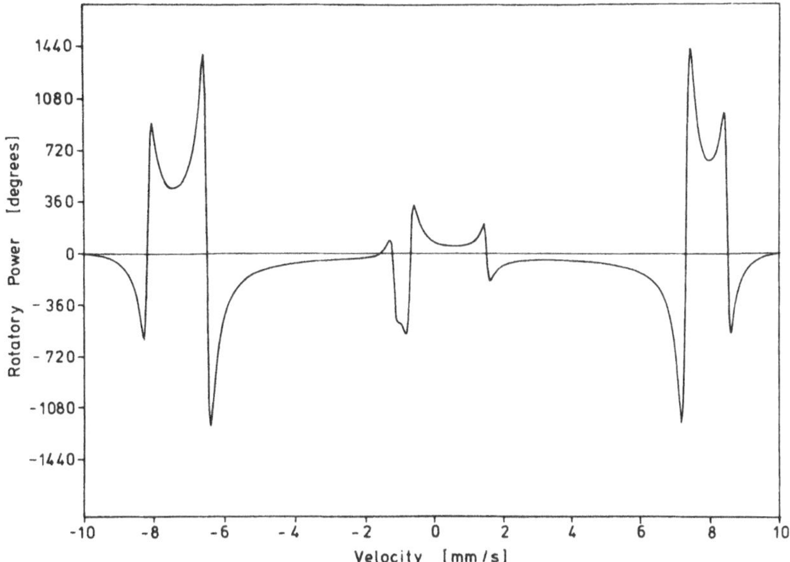


Fig. 6.19. Faraday rotatory power of the spinel $(Mg_{0.26};Fe_{0.74})[Mg_{0.74};Fe_{1.26}]O_4$ with a thickness of $\tau_T \approx 360$ in a longitudinal field ($\vartheta_T = 0°$), $H_T = 55$ kOe

the effect that the observed rotation, as given by (6.53), is mainly determined by the rotatory power at energies of the weakly attenuated source lines.

The stationary polarizer emits six linearly polarized lines. The four $\Delta m = \pm 1$ lines A, C, D, and F are at energies corresponding to -5.33, -0.83, +0.83, and 5.33 mm/s, respectively. Each of these may be considered to be a coherent superposition of two circularly polarized lines with opposite helicity. As can be seen from Fig.6.6, at least one helical component of each of these lines will be strongly absorbed in the transmission sample, while the component of opposite helicity will be transmitted. These surviving helically polarized lines cannot contribute to the measured Faraday rotation. On the other hand, the strong linearly polarized $\Delta m = 0$ lines B and E at ±3.08 mm/s are largely transmitted through the two "windows" which are nearly symmetrically located in the vicinity of ±4 mm/s. As seen in Fig.6.19 both of these lines are Faraday rotated in the same direction. However, one has to remember that the rotation represents contributions from both the octahedral and tetrahedral sites

$$\phi_{total} = [\phi] + (\phi) \quad .$$

Because the two sites are antiferromagnetically coupled, the rotations by the two sites are in opposite directions. In the experiment a net rotation of $19°$ was found. If one assumes that the $\Delta m = 0$ lines (B,E) are wholly responsible for the Faraday rotation, one can evaluate from the net rotation the relative populations of the [Fe] and (Fe) iron sites [6.8].

The sign of the magnetic hyperfine interaction can easily be found by comparing the sense of the Faraday rotation with the orientation of the external magnetic field at the transmitter [6.57].

It should be mentioned that IMBERT [6.17] in his pioneering work observed magnetic birefringence of linearly polarized $\gamma$ rays. With a double-motion Mössbauer spectrometer, he demonstrated that an "optical" axis can be induced in a transversely-magnetized enriched-$^{57}Fe$ transmission sample, thereby producing a phase shift between the ordinary and extraordinary component of the $\gamma$ ray. Using this technique, a "quarter-wave plate" can be realized. The change from linear to circular polarization in the transmitted $\gamma$ ray can be confirmed by the resulting invariance of line intensities in the rotating analyzer.

### 6.6.2 "Optical" Rotation

Finally, a description should be given of a search for Mössbauer "optical" activity in optically active chiral molecules containing $^{57}Fe$ [6.67]. From the classical theory of electromagnetic radiation one might expect an optical rotatory effect near resonances where the dispersion curves of left- and right-circularly polarized radiation may be different. A sample consisting of an optically active iron compound $(-)_{365}$-$C_5H_5Fe(CO)(I)(P)(C_6H_5)_2N(CH_3)CH(CH_3)(C_6H_5)$ enriched in $^{57}Fe$ was placed between a source (polarizer) of $^{57}Co$ in $\alpha$-Fe and an $\alpha$-Fe absorber (analyzer), both transversely magnetized ($\vartheta_A=\vartheta_S=90^\circ$) as shown schematically in Fig.6.18. The Mössbauer spectrum of the transmission sample is indicated in the center of the figure. The absorber rotates around the optical axis of the polarimeter. The source vibrates at $\approx 20$ Hz with a square curve form, moving alternately at two constant velocities ($v=\pm 8.41$ mm/s) to bring the source A and B or E and F lines into resonance with the absorber $\varepsilon$ and $\eta$, or $\alpha$ and $\beta$ lines, respectively. In the absence of a transmission sample, maximum resonant absorption for the linear polarization of source and absorber lines ($A\varepsilon$ and $B\eta$ or $E\alpha$ and $F\beta$) occurs when the orientation of $H_S$ and $H_A$ are perpendicular to each other ($H_S \perp H_A$, i.e., $\Delta\varphi=\varphi_A-\varphi_S=90^\circ$).

The detected $\gamma$ rays for $v = +8.41$ mm/s (right side of Fig.6.18) and $v = -8.41$ mm/s (left side of Fig.6.18) were accumulated in the two halves of a multichannel analyzer. The resulting Malus curves shown in the lower part of Fig.6.17 represent the $\gamma$ rays transmitted through the sample at positive-velocity (top) and negative-velocity (bottom) resonances. Fitting of these two Malus curves indicated a relative shift of $2.1^\circ$. However, this result is within experimental error.

The search for these rotatory effects was undertaken although it was realized that any positive result was rather doubtful. For the plane of linearly polarized radiation to be rotated by transmission through the sample, there must be some difference between the refractive indices for right and left circularly polarized radiation and therefore some net helicity of the $\gamma$-ray transition. But, as discussed in Sect.6.2.2, the electric field gradient of the iron sites does not completely lift the degeneracy of the nuclear levels (Fig.6.3), and only linear polarization (zero net helicity) is involved in each of the two possible γ-ray transitions. Therefore, rotation of the plane of linear polarization of γ radiation cannot occur under ordinary conditions. However, interference between photoelectric absorption and resonance absorption followed by internal conversion might result in a slightly different refractive index for right and left circular polarization. For visible light, optical activity is due to the interaction of the radiation with the "outer" electrons of the atoms which reflect by their polarizability the asymmetry of the optically active molecule or crystal. For high-energy γ rays, however, this would require an extremely high distortion of the inner K and L shell electrons by the optically active structure to obtain measurable effects. Mössbauer isotopes with E1 transitions (e.g., $^{181}$Ta) and high internal conversion coefficients might produce observable optical rotation of this type.

Two attempts have been made to observe optical activities with Mössbauer $^{57}$Fe polarimeters and transmission samples containing no $^{57}$Fe: one search was for "the optical rotation of quartz for 14.4-keV γ rays" [6.68], and the other for "the absorption of circularly polarized γ radiation in the L- and D-amino acids" [6.69].

*Acknowledgements.* A careful reading of the manuscript by Prof. R.S. Preston is gratefully acknowledged. The authors would also like to thank Dr. H. Spiering for his comments.

## References

6.1 M. Faraday: Phil. Trans. Roy. Soc. (London) *1* (1846)
6.2 W. Flagg, S.S. Hanna: Rev. Mod. Phys. *31*, 711 (1959)
6.3 R.L. Mössbauer: Z. Phys. *151*, 124 (1958)
6.4 J.G. Stevens, V.E. Stevens: *Mössbauer Effect Data Index* (Adam Hilger, London, and Plenum, New York, Washington, London )
6.5 E.P. Wigner: Rev. Mod. Phys. *29*, 255 (1957)
6.6 J.M. Daniels: *Oriented Nuclei-Polarized Target and Beams* (Pergamon, New York 1955) p.217
6.7 U. Fano: Rev. Mod. Phys. *29*, 74 (1957)
6.8 R.M. Housley, U. Gonser: Phys. Rev. *171*, 480 (1968)
6.9 R.M. Housley, R.W. Grant, U. Gonser: Phys. Rev. *178*, 514 (1969)
6.10 M. Blume, O. Kistner: Phys. Rev. *171*, 417 (1968)
6.11 S.S. Hanna, J. Heberle, C. Littlejohn, G.J. Perlow, R.S. Preston, D.H. Vincent: Phys. Rev. Lett. *4*, 177 (1960)
6.12 G.J. Perlow, S.S. Hanna, M. Hamermesh, C. Littlejohn, D.H. Vincent, R.S. Preston, J. Heberle: Phys. Rev. Lett. *4*, 74 (1960)
6.13 H. Frauenfelder, D.E. Nagle, R.D. Taylor, D.R.F. Cochran, W.M. Visscher: Phys. Rev. *126*, 1065 (1962)
6.14 H. Wegener, F.E. Obenshain: Z. Phys. *163*, 17 (1961)
6.15 H. Wiedersich: *Proceedings 2nd Symposium on Low-Energy X and Gamma Sources*, Vol.1, ORNL-11C-10, University of Texas (1967)
6.16 W. Keune, S.K. Date, I. Dézsi, U. Gonser: J. Appl. Phys. *46*, 3914 (1975)
6.17 P. Imbert: J. Phys. (Paris) *27*, 429 (1966)
6.18 R.M. Housley: Nucl. Instrum. Methods *62*, 321 (1968)

6.19 U. Gonser, H. Sakai, W. Keune: J. Phys. (Paris) *C6*, 709 (1976)
6.20 M. Henry, F. Varret: Phys. Status Solidi *A44*, 601 (1977)
6.21 J.P. Stampfel, P.A. Flinn: *Mössbauer Effect Methodology*, Vol.6, ed. by I.J. Gruverman (Plenum Press, New York, London 1971) p.95
6.22 S. Shtrikman: Solid State Commun. *5*, 701 (1967)
6.23 N.D. Heiman, J.C. Walker, L. Pfeiffer: Phys. Rev. *184*, 281 (1969)
6.24 N. Blum, L. Grodzins: Phys. Rev. *136*, A133 (1964)
6.25 C.E. Johnson, W. Marshall, J.G. Perlow: Phys. Rev. *126*, 1503 (1962)
6.26 U. Gonser, R.W. Grant, H. Wiedersich, S. Geller: Appl. Phys. Lett. *9*, 18 (1966)
6.27 R.W. Grant: In *Mössbauer Spectroscopy*, Topics in Applied Physics, Vol.6, ed. by U. Gonser (Springer, Berlin, Heidelberg, New York 1975) p.97
6.28 D. Barb: *Proc. Intern. Conf. Mössbauer Spectroscopy*, Vol.2 (Cracow, Poland 1975) p.379
6.29 M.T. Hirvonen: Nucl. Instrum. Methods *165*, 67 (1979)
6.30 M. Henry, F. Varret: Rev. Phys. Appl. *14*, 509 (1979)
6.31 R. Zimmermann: Nucl. Instrum. Methods *128*, 537 (1975)
6.32 H. Spiering: Habilitationsschrift, Universität Erlangen-Nürnberg (1978)
6.33 H. Willems, H. Fischer, A. Trautwein, U. Gonser, F. Parak, Y. Maeda: J. Phys. (Paris) *C1*, 487 (1980)
6.34 U. Gonser, R.W. Grant: Phys. Status Solidi *21*, 331 (1967)
6.35 R.W. Grant, S. Geller, H. Wiedersich, U. Gonser, C.D. Fullmer: J. Appl. Phys. *39*, 1122 (1968)
6.36 S. Geller, R.W. Grant, L.D. Fullmer: J. Phys. Chem. Solids *31*, 793 (1970)
6.37 U. Gonser: In *Hyperfine Structure and Nuclear Radiations*, ed. by E. Matthias, D.A. Shirley (North Holland, Amsterdam 1968) p.343
6.38 U. Gonser, R.M. Housley, R.W. Grant: Phys. Lett. *29*, 36 (1969)
6.39 J.B. Forsyth, I.G. Hedley, C.E. Johnson: J. Phys. C (Proc. Phys. Soc.) *1*, 179 (1968)
6.40 C.E. Johnson: J. Phys. (Paris) *C2*, 1996 (1969)
6.41 R.W. Grant, S. Geller: Solid State Commun. *7*, 1291 (1969)
6.42 R.W. Grant, H. Wiedersich, R.M. Housley, G.P. Espinosa, J.O. Artman: Phys. Rev. *B3*, 678 (1971)
6.43 R.W. Grant: Phys. Rev. *42*, 1619 (1971)
6.44 S. Morup: J. Phys. (Paris) *35*, C6-683 (1974)
6.45 T.C. Gibb: J. Phys. C *8*, 229 (1975)
6.46 T.C. Gibb: Chem. Phys. Lett. *30*, 137 (1975)
6.47 H. Fischer, U. Gonser, H.D. Pfannes, T. Shinjo: *Proc. Int. Conf. Mössbauer Spectroscopy*, Cracow, Poland *1*, 463 (1975)
6.48 H.-D. Pfannes, H. Fischer: Appl. Phys. *13*, 317 (1977)
6.49 T. Shinjo, H.D. Pfannes, U. Gonser: *Proc. Int. Conf. Mössbauer Spectroscopy*, Cracow, Poland *1*, 465 (1975)
6.50 M.T. Hirvonen, A.P. Jauho, T.E. Katila, K.J. Riski, J.M. Daniels: Phys. Rev. *B15*, 1445 (1977)
6.51 U. Gonser, M. Ghafari, H.G. Wagner: J. Magn. Magn. Mater. *8*, 175 (1978)
6.52 J.D. Bernal: Nature *185*, 68 (1960)
6.53 H. Fischer, U. Gonser, R.S. Preston, H.G. Wagner: J. Magn. Magn. Mater. *9*, 336 (1978)
6.54 U. Gonser, H.D. Pfannes: J. Phys. (Paris) *C6*, 113 (1974)
6.55 H.D. Pfannes, U. Gonser: Nucl. Instrum. Methods *114*, 297 (1974)
6.56 D. Barb, G. Giolu, M. Rogalski: *Proc. Int. Conf. Mössbauer Spectroscopy*, Bucharest, Romania *1*, 21 (1977)
6.57 U. Gonser, R.M. Housley: Phys. Lett. *26A*, 157 (1968)
6.58 J.P. Hannon, G.T. Trammell: Phys. Rev. *186*, 306 (1969)
6.59 Y.M. Aivazyan, V.A. Belyakov: Sov. Phys.-Solid State *13*, 808 (1971)
6.60 A.V. Mitin: Phys. Status Solidi *B53*, 93 (1972)
6.61 J.P. Hannon, N.J. Carron, G.T. Trammell: Phys. Rev. *B9*, 2791, 2810 (1974)
6.62 U. Gonser, H.D. Pfannes: *Proc. Int. Conf. Magnetism*, Moscow, USSR *3*, 117 (1974)
6.63 V.G. Labuskin, S.N. Ivanov, G.V. Chechin: JETP Lett. *20*, 157 (1974)
6.64 A.V. Mitin, G.P. Chugonova: Sov. Phys.-Solid State *16*, 403 (1974)
6.65 D. Barb, M. Rogalski: *Proc. Int. Conf. Mössbauer Spectroscopy*, Cracow, Poland (1975) p.469
6.66 A. Kastler: C.R. Acad. Sci. *255*, 3397 (1962)

6.67 U. Gonser, H. Engelmann, H. Brunner, M. Mushiol, W. Nowak: In *Workshop on New Directions in Mössbauer Spectroscopy*, ed. by G.J. Perlow, AIP Conf. Proc. *38*, 87 (1977)
6.68 L. Grodzins, J. Alonso: Rev. Mod. Phys. *36*, 359 (1964)
6.69 L. Keszthelyi, I. Vincze: *Proc. Int. Conf. Mössbauer Spectroscopy*, Cracow, Poland *1*, 447 (1975)

# 7. Iron-Ion Implantation Studied by Conversion-Electron Mössbauer Spectroscopy

B. D. Sawicka and J. A. Sawicki

With 19 Figures

Ion implantation has in recent years become a subject of considerable interest and great use in materials science and technology. The technique allows the introduction of any atoms at well-controlled rates into almost any kind of solids, regardless of usual thermodynamic constraints. Therefore, the production of a variety of materials with new and sometimes surprising properties is possible. As a tool of a well-controlled doping, ion beams are already successfully used in the manufacturing of semiconductor components. Ion implantation is also utilized for investigating a variety of phenomena in metals including nonequilibrium alloying, radiation damage, surface modifications, erosion and wear, corrosion, etc.

Mössbauer spectroscopy has proved very useful in studies of ion implantation and properties of implanted materials. Hyperfine interactions measured by this method provide information on the s-electron density, the electric field gradient, and the magnetic field at the implanted probe nucleus. The vibrational behavior and the dynamics of implanted atoms can also be followed. From this, much can be deduced about the electronic structure of implanted atoms, about their position in the host matrix and local environment, about the annealing of lattice defects, about aggregation processes, etc.

Starting from about 1965 many interesting results were obtained by Mössbauer spectroscopy studies of radioactive ions implanted either as the reaction or decay recoils or by the use of isotope separators. Various experimental techniques utilizing radioactive implants in investigations of hyperfine interactions were reviewed by DE WAARD [7.1,2].

Implantation of non-radioactive ions, which is sometimes more advantageous, can be very efficiently studied by means of the conversion electron Mössbauer spectroscopy (CEMS). This field has been successfully developed in Cracow since 1973 and is now studied also in several other laboratories. So far the implantation of $^{57}$Fe has been mostly employed but there are prospects of expanding the application of this method to several other Mössbauer isotopes.

In this chapter we first discuss shortly the experimental aspects of iron implantation, the properties of implanted materials and the applications of CEMS technique in their studies. Next, representative results for $^{57}$Fe implanted metals, semiconductors and insulators are presented. More extensive survey of the experimental technique and the results can be found in works [7.3,4] and in references of Sections 7.2-4.

## 7.1 Implantation of Iron Ions

Ion implantation is a process of introducing atoms in the form of a beam of energetic ionized particles into a substance. Ions can be introduced into a host matrix in well-defined and large quantities, even orders of magnitude above the equilibrium solubility limits. The depth to which the material is doped is determined by the energy of projectiles and by the ratio between atomic masses of the projectile and the host atoms. The upper concentration of implanted impurities and their depth distribution is, however, influenced also by the sputtering effects. The depth distribution, sputtering yields, and lattice damage are important factors in ion implantation.

There are very good introductions to the technique of ion implantation and the associated physical phenomena, e.g., [7.5-8]. Rapid progress in the applications of ion implantation can also be followed from proceedings of recent conferences, e.g., [7.9,10]. In this section we present some basic information concerning isotope separator implantation of $^{57}$Fe.

### 7.1.1 Stopping and Range of Ions

The theoretical description of the slowing-down process of energetic ions in solids is based on the work of LINDHARD, SCHARFF and SCHIØTT (LSS theory) [7.11,12]. Low-energy ions are mainly stopped through elastic collisions with individual screened nuclei in the target (nuclear stopping). High-energy ions lose their energy mainly via inelastic collisions in which electrons are excited or rejected from the interacting atoms (electronic stopping). For iron ions, with energies below about 50-100 keV, the slowing-down is dominated by the nuclear stopping.

The depth distribution of atoms implanted in an amorphous target can be well-approximated by a Gaussian. The concentration of the implanted atoms vs depth R can be expressed as

$$x(R) = x_{max} \exp[-(R-R_p)^2/2\Delta R_p^2] \quad ,$$

where $R_p$ is the average projected range along the incident beam direction, $\Delta R_p$ is the straggling in the projected range, and $x_{max}$ is the concentration at depth $R_p$. The average concentration of implanted atoms (the fractional content of implanted impurities related to the number of host atoms), $\bar{x}$, can be defined as a ratio between the number of implanted atoms to the number of host atoms in a layer $4\Delta R_p$. With such a definition $\bar{x}$ equals $\bar{x} = 0.62x_{max} = 0.42AD/\Delta R_p$, where A is the mass number of host atoms, and in order to get $\bar{x}$ in percent, the number of implanted atoms per unit area D (dose) is given in units of $10^{16}/cm^2$ and $\Delta R_p$ is in $\mu g/cm^2$.

In Fig.7.1 the values of $R_p$, $\Delta R_p$ and $\bar{x}$ for Fe implants in Al host, calculated using the SCHIØTT approximation rules [7.13], are plotted vs ion energy. Similar graphs for iron in some other hosts are also available [7.4]. In Fig.7.2 the values

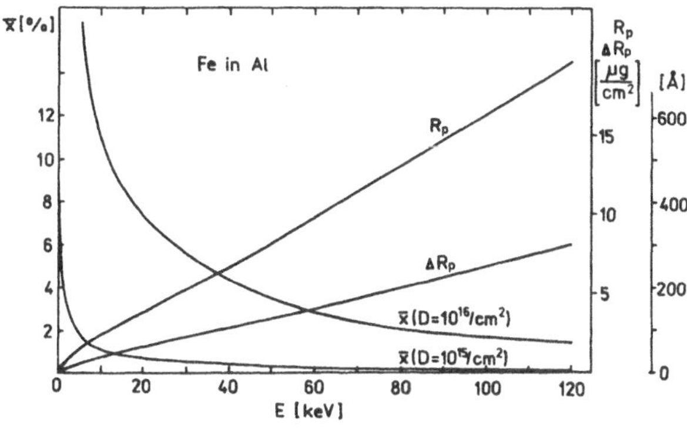


Fig. 7.1. Range $R_p$, range straggling $\Delta R_p$ and average ion concentration $\bar{x}$ for a total dose of $10^{15}$ and $10^{16}$ $^{57}Fe/cm^2$ in an aluminum host as a function of incident ion energy [7.28]

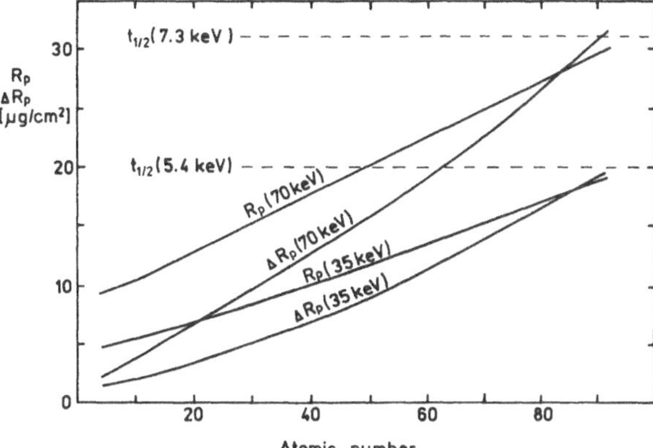


Fig. 7.2. Range $R_p$ and range straggling $\Delta R_p$ for iron implants at two incident ion energies plotted vs the atomic number of the host element. Dotted lines present approximate half-absorption thicknesses $t_{1/2}$ for 7.3-keV and 5.4-keV electrons and were obtained assuming exponential attenuation of electrons [7.3]

of $R_p$ and $\Delta R_p$ for Fe implants are plotted vs the atomic number of the host element. Values of $R_p$ and $\Delta R_p$ (in units of a mass per area) increase with the increasing atomic number of the host element. For instance, for 70-keV Fe ions in Al one has $R_p$ = 11.2 μg/cm² (415 Å) and $\Delta R_p$ = 4.6 μg/cm² (170 Å), while in Au $R_p = \Delta R_p$ = 27 μg/cm² (140 Å). The average atomic concentration $\bar{x}$ at a given dose D depends only slightly on the atomic number of the host, especially for heavy elements. For instance, for $10^{16}$ of 70-keV $^{57}Fe$ implants per cm², one has $\bar{x} \simeq 2.4\%$ for an Al target, $\bar{x} \simeq 2.8\%$ for Ti, and $\bar{x} \simeq 3\%$ for heavier targets. Experimental ranges are in a fairly good agreement with the ranges predicted by LSS theory if channeling and diffusion of ions can be neglected.

The energy and momentum imparted to target atoms in primary collisions is dissipated via secondary collisions and displacements. A radius of a collision cascade is typically about 100 Å. The number of displaced atoms N(E) depends on the ion energy E, and the effective displacement energy $E_d$, $N(E) \sim 0.4/E_d$. Since the value of $E_d$ is about 15 eV for semiconductors and 20-40 eV for metals, the number of displaced atoms for 100-keV projectile is more than 1000.

Displacement collisions result both in damaging and sputtering of the host material. The sputtering yield Y determines the average number of atoms ejected from the

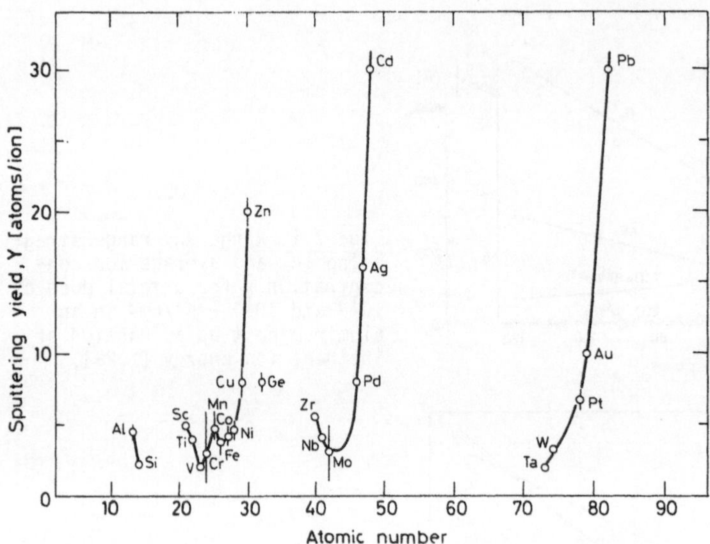


Fig. 7.3. Sputtering yields for 50-keV $Fe^+$ bombardment at room temperature [7.14]

host surface per one incident ion. The sputtering can significantly alter the distribution of implants with depth and limits the high doping possible by the implantation technique. It is estimated that the maximum concentration of implanted atoms cannot exceed, approximately, the reciprocal of the sputtering yield, $x_{max}\% \lesssim 1/Y$.

Comprehensive reviews of sputtering phenomena are given in [7.5,14]. The sputtering depends on the atomic properties of both the target and the incident ions. Generally, the value of sputtering yield increases with ion energy and saturates at some energy, which in the case of $Fe^+$ bombardment is expected to be about 100 keV. Information on sputtering yields has been acquired for a variety of ion-target combinations, but there are only very few experimental values for iron projectiles. The data in Fig. 7.3, taken from [7.14], were obtained mostly by interpolation between measured and calculated values for 50-keV $Ar^+$ and $Kr^+$ bombardment at room temperature at normal incident angle. The data indicate that iron can be implanted in most metals up to fairly high concentrations, though in some cases such as Zn, Ag, Cd and Pb, the sputtering may markedly reduce the expected final concentration of iron. For studies of Fe implants with the use of the isotope separator it is practical to keep in mind that the sputtering corrections for determination of $\bar{x}$ are fairly small for Y smaller than about 5 and doses D smaller than about $10^{16}$ atoms/$cm^2$.

### 7.1.2 Implanted Materials

Ion implantation is characterized by a large number of interstitials and vacancies created in collision cascades and by a rapid quenching of damaged zones. It is estimated that the collision cascade is completed in about $10^{-13}$ s, whereas the quenching rate is in the range of $10^{10}$-$10^{15}$ K/s [7.8]. Therefore, some parallels established between materials produced by ion implantation and those obtained by vapor or liquid quenching techniques are not surprising.

The properties of ion-implanted materials strongly depend on the mobility of defects, and thus the temperature of the target during implantation is an important factor. Ion implantation at low temperatures makes possible creating and then studying various defects (such as impurity-vacancy systems which are mobile or unstable at elevated temperatures) and then to observe their annealing stages. Mössbauer studies of lattice defects are usually performed with radioactive nuclei implanted at relatively small doses of about $10^{12}$ atoms/cm$^2$. Applications of Mössbauer spectroscopy in investigations of defects and radiation damage phenomena were discussed by GONSER [7.15], VOGL [7.16,17], DÉZSI [7.18] and SAWICKA [7.19].

When metals are implanted at room temperature, many damage sites anneal out in a very short time mostly by recombination of closely spaced vacancies and interstitials, while the processes of migration of implanted impurities are usually not very efficient. Therefore, room temperature ion implantation in metals can preferably lead to the creation of metastable and mostly crystalline phases. In some cases the aggregation of impurity atoms can be observed occurring at its very beginning, as a result of short-range order processes. Diffusion as well as further aggregation of impurities into larger clusters and the precipitation of stable alloy phases is possible at elevated temperatures. Some results of studies of metastable phases and iron aggregation processes in metals will be discussed in Sect.7.3.1,2.

The location of implanted atoms in the host matrix depends on many factors, the relative sizes of the implanted and host atoms being a significant one. The characterization of sites occupied by iron impurities in various metals on the basis of their Mössbauer spectra will be discussed in Sect.7.3.2.

Under implantation, solids exhibiting strong covalent bonds behave differently than metals. Their crystal structure can easily be destroyed by ion implantation at room temperature. In silicon and germanium, a transition from crystalline to amorphous phase due to the overlap of individual damaged zones takes place at doses of about $10^{14}$ atoms/cm$^2$. The nature of a new phase and the role played by impurities implanted in the random network is not yet clear. Studies of iron implanted in silicon and germanium at doses exceeding the amorphisation doses will be summarized in Sect.7.3.3.

Over the past few years there has been also a rapidly increasing interest in the study and application of ion implantation in insulators such as, e.g., glasses, ceramics, refractory oxides, silicates, nitrides, carbides and ionic crystals. The mechanism of ion stopping and the nature of damaged regions in such materials is generally more complex than in elemental hosts. The phenomena related to chemistry of hot atoms play a substantial role. The implantation-induced formation of non-equilibrium molecular clusters or the nature of complex impurity-vacancy phases can be studied. The applications of nuclear techniques, and of the Mössbauer spectroscopy in particular, in this area of materials science have been rather seldom so far. Some examples of Mössbauer effect studies of $^{57}$Fe in ionic crystals and oxides will be quoted in Sect.7.3.4.

### 7.1.3 Mössbauer Study of Iron Implanted Materials

The Mössbauer spectroscopic studies of implanted iron impurities were initiated in 1965 by SPROUSE et al. [7.20]. They used recoil implantation of Coulomb-excited $^{57}$Fe and measured, on-beam, the emission spectra of the 14.4-keV transition. Doses studied by this method are as small as $10^{10}$-$10^{12}$ atoms/cm$^2$. Ions are implanted with a high energy (~30 MeV). The state of atoms in a very short time after implantation (~100 ns) is observed. The technique of recoil implantation of $^{57}$Fe is rarely employed as it requires a long working time of the accelerator. One has also to realize that the dose determination is difficult in these experiments (the dose accumulates in the course of the measurements).

The isotope-separator implantation of radioactive parent $^{57}$Co isotope for recording the $^{57}$Fe Mössbauer spectra was initiated in 1970 by DE BARROS et al. [7.21]. This technique made it possible to study a range of doses of about $10^{12}$-$10^{14}$ atoms/cm$^2$ and therefore concentrations of $10^{-4}$-$10^{-2}$ at.%. The method is especially useful when studying the interaction between impurities and point defects and the consecutive annealing stages. The examples of such studies can be investigation of $^{57}$Co in Al at low temperatures [7.22,23] and $^{57}$Co in Si [7.24]. A long-lived radioactive contamination of the isotope-separator ($T_{1/2}$=270 d) and small efficiency of implantation (1-5%) are disadvantages of this method. Moreover, aftereffects of $^{57}$Co decay may complicate the analysis of the spectra, especially in insulators.

First studies of isotope-separator implanted stable $^{57}$Fe were reported in 1973 by SAWICKI et al. [7.25-27]. In initial experiments the transmission spectra were measured for samples implanted at high doses, $10^{16}$-$10^{17}$ atoms/cm$^2$. Since 1975 the conversion-electron Mössbauer spectroscopy (CEMS) was applied [7.28-31] which made it possible to study also much smaller doses. Many papers were published since then reporting the $^{57}$Fe implantation CEMS study (see references in Sect.7.3).

Doses measured by CEMS are, at present, $5\cdot10^{13}$-$10^{17}$ of $^{57}$Fe atoms/cm$^2$. Concentrations of implanted impurities are in a range from $10^{-2}$ at.% up to tens of atomic percent. In special cases the dose as low as $10^{13}$ atoms/cm$^2$ can be detected, but in principle doses lower than $10^{14}$ atoms/cm$^2$ require the $^{57}$Co implantation and the emission spectra measurements.

### 7.1.4 Technique of $^{57}$Fe Implantation

Because of the small abundance of $^{57}$Fe (2.14%) compared to $^{56}$Fe (91.7%), a magnetic mass separator with a sufficient dispersion is required for performing the $^{57}$Fe implantation. It is also convenient to use an enriched initial material, e.g., anhydrous $FeCl_2$ enriched in $^{57}Fe$ to about 15-30%. In the isotope separator used in Cracow, $Fe^+$ ions are produced in an efficient slit type ion source. The final energy of ions may be varied in a range 5-80 keV. For $^{57}Fe$ implantation, currents of iron ions on the target have always been smaller than 3 μA/cm$^2$. An estimated admixture of $^{56}Fe$ has been smaller than 5%. The position of the beam on the target is controlled with wire probes. A good homogeneity and a large area of implantation are ensured by sweeping a target holder, with about 20 samples having an area of 1 cm$^2$ each being implanted simultaneously. An implantation with a dose of $10^{16}$ atoms/cm$^2$ lasts less than 10 hours. A special treatment of the target just before or immediately after implantation is possible. For instance, the targets can be cleaned by sputtering using a low-energy $Ar^+$-beam or a protective layer can be deposited on the sample before taking it out of the separator vacuum chamber in order to protect the surface against the oxidation. More details about the $^{57}Fe$ implantation technique can be found in [7.4,32].

## 7.2 Conversion-Electron Mössbauer Spectroscopy

### 7.2.1 The Method

The recoilless resonant absorption of gamma rays can be measured in transmission or scattering geometry. The latter technique has many advantages, but because it is more sophisticated, it is used less commonly. In Mössbauer scattering experiments one measures either the reemission of gamma rays or detects electrons or X rays emitted from the resonant sample due to the internal conversion process. A choice of the radiation detected determines the thickness of the surface layers to be examined. A comprehensive discussion of various scattering experiments in Mössbauer spectroscopy was presented by WAGNER [7.33] and more recently by CHAMPENEY [7.34].

In conversion electron Mössbauer spectroscopy (CEMS) the spectra of recoilless resonant absorption are registered via detection of low-energy electrons. The essential feature of CEMS is that it makes possible an analysis in situ and in a nondestructive way of the shallow surface layers of solids of 1-100 nm. Next, in favorable cases, CEMS is the most sensitive technique on comparison to other types of Mössbauer measurements.

Because of its advantageous features the 14.4-keV transition in $^{57}Fe$ is the most commonly used in Mössbauer spectroscopy. This transition is also especially suitable for conversion electron Mössbauer measurements. Data on electrons and photons reemitted after resonant excitation of the 14.4-keV state of $^{57}Fe$ are given in Table 7.1. The data indicate that 7.3-keV K-shell conversion electrons and 5.4-keV K-LL Auger electrons can be used to probe layers as thin as 10-$10^2$ nm, whereas layers of a few micrometers can be examined with backscattered gamma or X rays.

Table 7.1. Energies, intensities, and ranges of photons and electrons emitted after resonant excitation of the 14.4-keV state of $^{57}$Fe (electron conversion coefficient $\alpha$=8.2). Radiations emitted with energies lower than 0.5 keV are not included. The intensities are given per decay of an excited state. The ranges for photons are the mean penetration depths for photoelectric attenuation. Upper and lower limits given for the electron ranges are the maximum range and the mean free path for penetration without energy loss, respectively (adopted from [7.33])

| Type of emitted radiation | E [keV] | Intensity | Range r in Fe metal |
|---|---|---|---|
| gamma-rays | 14.4 | 0.11 | $r \approx 20$ µm |
| K X-rays | 6.3 | 0.28 | $r \approx 20$ µm |
| L X-rays | 0.7 | 0.002 | |
| K-shell conversion electrons | 7.3 | 0.79 | 10 nm $\lesssim r \lesssim$ 400 nm |
| L-shell conversion electrons | 13.6 | 0.08 | 20 nm $\lesssim r \lesssim$ 1.3 µm |
| M-shell conversion electrons | 14.3 | 0.01 | 20 nm $\lesssim r \lesssim$ 1.5 µm |
| K-LL Auger electrons | 5.5 | 0.6 | 7 nm $\lesssim r \lesssim$ 200 nm |
| L-MM Auger electrons | 0.5 | 0.6 | 1 nm $\lesssim r \lesssim$ 2 nm |

The sensitivity of the CEMS method is connected with very high signal-to-noise ratio which can be obtained in CEMS measurements. The CEMS technique is advantageous when the photoelectric attenuation of gamma rays in a resonant scatterer is small and when the detection of electrons can be carried out at large solid angle, cf. e.g. [7.35,36]. The first condition requires a high value of the ratio between the resonant and photoelectric cross sections. For $^{57}$Fe this value is high, $\sigma/\sigma_{ph} = 300$. The second condition is fulfilled by a proper choice of electron detectors (see Sect.7.2.3).

In the last decade the CEMS technique has been much exploited for studies of a variety of phenomena such as corrosion, oxidation, surface magnetism, ion implantation, etc. Experiments are mostly performed using $^{57}$Fe, but also $^{119}$Sn, $^{181}$Ta, $^{151}$Eu and $^{169}$Tm were practically applied in some investigations of surface phenomena. Examples and references for further reading can be found in surveys by SPIJKERMAN [7.37], WEYER [7.38] and TRICKER [7.39].

### 7.2.2 CEMS of $^{57}$Fe Implants

Because of its high sensitivity in comparison to other types of Mössbauer measurements and because of its ability to detect resonant atoms deposited only at surface layers, the CEMS method is especially adequate in studying the implanted samples.

First, practically only the resonant atoms which are situated in the outermost layers of materials contribute to the CEMS spectrum. The depth of these layers determined by the range of conversion electrons is close to the range at which the implanted atoms are deposited. In Fig.7.2 the range $R_p$ or iron implants in solids is compared to the half-absorption thickness $t_{1/2}$ for attenuation of 7.3-keV and 5.4-keV electrons. The corresponding half-absorption thicknesses were obtained in a

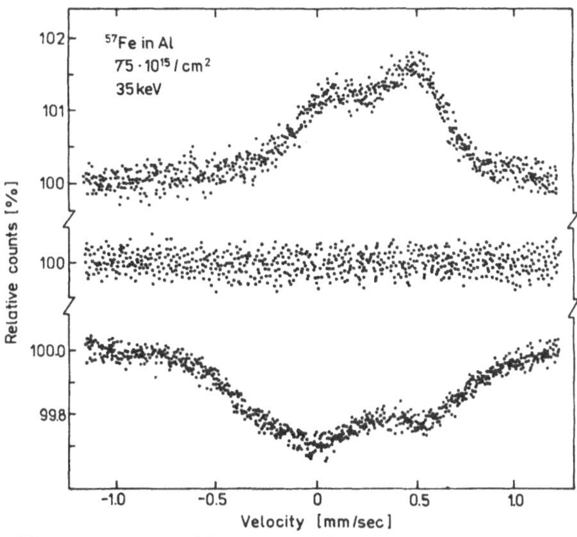


Fig. 7.4. CEMS spectrum of $^{57}$Fe-implanted aluminum (top) and transmission Mössbauer spectrum (bottom) of the same sample ($^{57}$Fe dose of $10^{15}$ atoms/cm$^2$, ion energy 35 keV, $\bar{x} \sim 0.5\%$, foil thickness ~0.3 mm). CEMS spectrum was measured in one day, transmission spectrum in ten days. CEMS measurement of nonimplanted foil (middle) indicates nil effect. Iron impurities in the bulk of a target foil are observed only in transmission spectrum as a broad line near zero velocity [7.28]

rough approximation of an exponential attenuation of electrons, using data on electron transmission for Al, Cu, Ag and Au in the energy range of 5-15 keV [7.3,40]. As seen, the $R_p$ values are mostly smaller than the $t_{1/2}$ values. More precise calculations, based on the KRAKOWSKI and MÜLLER semiempirical approach [7.41] or on a Monte Carlo technique of LILJEQUIST et al. [7.42] also indicate that electrons are emitted from implanted layers with very high efficiency and small energy losses.

Next, the sensitivity of CEMS technique, superior to other types of Mössbauer measurements, is an important feature in application to iron implantation. The transmission measurements were performed for doses of $10^{16}$ $^{57}$Fe/cm$^2$ but this is a practical limit of this technique. The CEMS measurements for doses higher than about $10^{15}$ $^{57}$Fe/cm$^2$ presents no problem in all targets. In most favorable cases, e.g., for Fe in Al and Si, doses as small as $10^{14}$ atoms/cm$^2$ or even $5 \cdot 10^{13}$ atoms/cm$^2$ were also used [7.30,32,43]. It is to be noted that $10^{14}$ of Fe atoms per cm$^2$ is equivalent to less than 10 ng/cm$^2$ or to about 1/20 of a monoatomic layer !

At last, samples of any shape and any thickness can be used for CEMS. This is the advantage in comparison to transmission geometry, for which the conditions of attenuation of the gamma-rays set the limits on the sample thickness. With the exception of targets of very light elements, the thicknesses no higher than a few mg/cm$^2$ can be applied in transmission measurements. This is frequently difficult to achieve and to handle, especially for implanted samples. Furthermore, since samples are usually much thicker than the depth implanted, the traces of unwanted resonant impurities diluted in the bulk can amount to the dose of implanted atoms (iron residual impurities are frequently difficult to avoid). The bulk impurities may contribute to spectra in transmission but they are not detected in CEMS spectra.

The usefulness of CEMS in studies of $^{57}$Fe implantation was pointed out by SAWICKI et al. in 1975 [7.28-30]; an example of their first results is shown in Fig.7.4. Since then many experiments on $^{57}$Fe-implanted samples have been carried out, mainly for samples implanted with doses of $10^{15}$-$10^{16}$ atoms/cm$^2$, as presented in next sections.

### 7.2.3 Experimental Technique

The helium-flow proportional backscatterer counters have so far been most useful in CEMS studies of $^{57}$Fe-implanted samples. Such counters are virtually insensitive to incident gamma and X rays and may detect electrons at a large solid angle and with high efficiency. Many counters of this type were described in the literature [7.35, 36]. The fabrication of the counter for a room temperature work and its handling is easy.

An example of a counter used for room temperature measurements is shown in Fig.7.5. A counter body (Fig.7.5) was made of a lucite box and a two-wire anode was installed

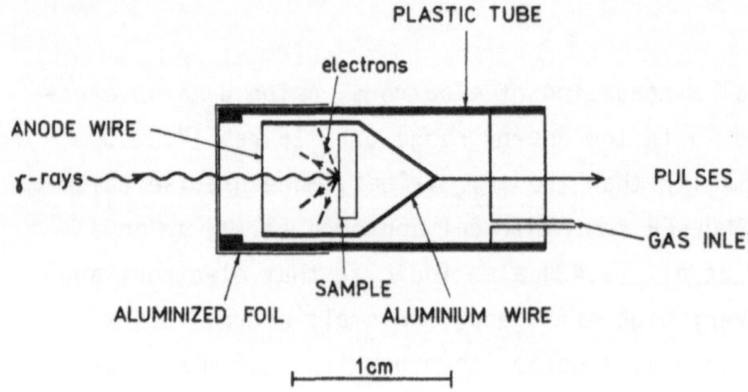


Fig. 7.5. Miniature, helium-flow counter constructed for room temperature CEMS measurements in the author's laboratory. This counter was also used in CEMS experiments under high external magnetic fields [7.3]

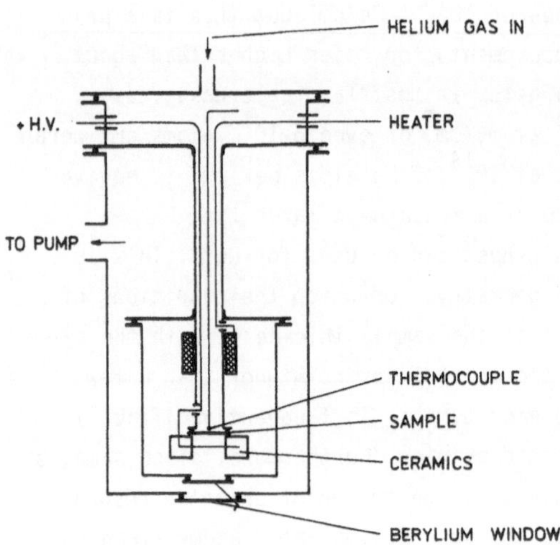


Fig. 7.6. Proportional counter constructed for CEMS measurements at high temperatures. Pure helium was used as working gas [7.50]

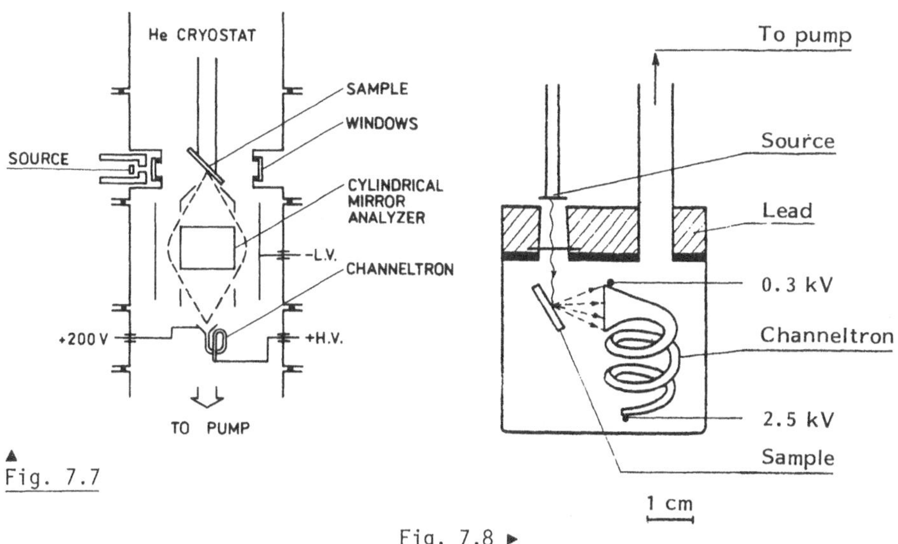


Fig. 7.7. Typical experimental equipment for CEMS measurements with channeltron. The unit is attached to the cryostat and can be used for temperature measurements. Electrons rejected from the sample are focused at the channeltron entrance by means of the electrostatic mirror analyzer. Option with retarding grids for measurements of low energy electrons, below 100 eV, was also used in the author's laboratory [7.53]

Fig. 7.8. A simple unit for CEMS measurements with channeltron which can be used for temperature measurements

perpendicularly to the direction of the gamma rays. The flow of helium gas during the work of the counter was maintained at the rate of about 1 $cm^3$/min. The anodic potential was about 700 V. When operating with a helium-methane mixture or pure helium, such detectors are insensitive to gamma rays, therefore the sample can be simply placed against the counter back wall or with the sample surface closing an opening in the back wall. In order to avoid a high background of photoelectrons arising from interaction of the incident gamma or X rays with the constructive parts of the counter, these should be made of materials with low atomic numbers.

The counter presented in Fig.7.5 was designed to be used for CEMS measurements under an external magnetic field [7.44]. The shape and size of the counter allowed for its convenient installation in a superconducting solenoid. These measurements were performed with the sample (scatterer) placed inside of the counter and the magnetic field up to 10 T, applied parallel to the direction of the gamma rays [7.44], see also Sect.7.3.3. To allow the location of a source at zero field at small distance from the scatterer in a high field, an additional compensating superconducting coil was used. In this way a sufficiently high count rate of the order of 1000 pulses/s for a 10-mCi $^{57}$Co source was achieved.

With the use of various helium counters, CEMS measurements were made at various temperatures, so far within 30 K to 600 K. For such measurements a counter must be installed in a cryostat or furnace, as described by ISOZUMI et al. [7.45,46] and SAWICKI et al. [7.47-50]. An example of equipment used for measurements at high temperatures is shown in Fig.7.6.

Various other detectors of low-energy electrons, like parallel-plate avalanche counters [7.38], channeltrons [7.51], plastic scintillators can also be applied in CEMS measurements. Channeltrons are of particular interest because of their high efficiency for registration of low-energy electrons (90% for electrons below 500 eV and 10% for 10 keV electrons) as well as because of a very high gain (about $10^7$-$10^8$) and a low dark current. Channeltrons (and multichannel plates) offer a possibility of CEMS measurements at variable temperatures and can be combined with a variety of cryostats and vacuum systems. However, because of the small size of these detectors and the necessity of screening against photons, the solid angle in CEMS measurements with channeltron is rather small.

An equipment for CEMS measurements at various temperatures from 4.2 K with channeltron was reported by MASSENET [7.52] and by TYLISZCZAK et al. [7.53]. In both cases the channeltron was kept at room temperature while the sample was attached to the cryostat (Fig.7.7). A much simpler setup can be designed when using a special type of channeltron which can operate efficiently in a wide range of temperatures. With the unit constructed recently (Fig.7.8), a very high efficiency was obtained, as good as that for a gas detector. This unit can be immersed into a liquid nitrogen or liquid helium dewar, which greatly simplifies the equipment and measurement procedure. A simple construction of the experimental setup in order to avoid a high nonresonant background (similarly as for gas detectors) and a good collimation of the incident gamma rays are important to get a good signal-to-noise ratio.

The depth-selective CEMS requires analyzing electron spectra with a good energy resolution [7.54]. Such experiments must be performed using electron spectrometers with high luminosity and high transmission. In spite of a considerable progress which was made in recent years both in experimental technique and in data analysis (see, e.g., [7.55,56] for references), the depth resolution of the method is still not sufficient for a precise determination of ion implantation profiles. Studies of $^{57}$Fe-implanted aluminum with depth-selective CEMS were recently initiated by JONES et al. [7.57].

## 7.3 Studies of Iron Implanted in Various Hosts

The experimental technique presented above has already been used in studies of $^{57}$Fe-implanted metals, semiconductors, and insulators. The main results obtained for iron in aluminum, metals of the 3d, 4d and 5d series, as well as for silicon, germanium and some insulators will be summarized below. Studies in Cracow were done for the isotope-separator implantation of $^{57}$Fe at room temperature and with ion energy varied from 10 to 70 keV for different samples.

### 7.3.1 Aluminum

The range of doses of $^{57}$Fe ions implanted in aluminum for CEMS study was from $10^{14}$ to $2\times10^{17}$ atoms/cm$^2$ [7.28,32]. The resultant iron concentration $\bar{x}$ varied from 0.02% to about 30% in various samples, which was always above the solid solubility limit of Fe in Al at equilibrium conditions (0.001% at room temperature and 0.025% at 650°C). Examples of the spectra are shown in Figs.7.9,10. Spectra indicated as systematic change with the average iron concentration $\bar{x}$ in the sample and independently on whether various $\bar{x}$ was achieved by changing iron dose or energy.

At iron concentrations $\bar{x}$ smaller than 5%, the spectra are composed of a single line and a quadrupole-split doublet. The contribution of the single line increases with the decrease in $\bar{x}$, as shown in Fig.7.11. The single line has been ascribed to iron atoms having no nearest Fe neighbors (iron monomers). The doublet was ascribed to iron aggregates (dimers mostly). The values of the isomer shifts for both components show that the s-electron density at nuclei in iron dimers is larger than in iron monomers and smaller than in metallic iron (for $\alpha$-Fe $\delta$=0 mm/s). It is to be noted that similar results were obtained by NASU et al. [7.58]. These authors studied splat-quenched FeAl alloys from 0.5 to 5 at.% Fe by the transmission technique and they also proved the quadrupole character of the doublet by measuring the spectra in a high external magnetic field.

In random solid solutions the probability of monomers, dimers, etc. is given by binomial distribution. For instance, the probability P(n) that an Fe atom in the fcc matrix with an iron concentration $\bar{x}$ has n Fe atoms as nearest neighbors is given by

$$P(n) = \binom{12}{n} x^n (1-x)^{12-n} \quad , \text{ where } \quad n = 0,1,2,\ldots,12 \quad .$$

For small x, iron monomers and dimers are most probable, e.g., for $x = 1\%$, $P(0) = 0.88$ and $P(1) = 0.11$ and for $x = 5\%$ $P(0) = 0.53$ and $P(1) = 0.34$. The comparison of the fractional intensity of the single line in the spectra for Fe in Al with the probability of monomers P(0) in fcc matrix (Fig.7.11) indicates the enhanced formation of dimers in implanted samples, especially at high iron concentrations.

Iron aggregation processes can be strongly enhanced both by implantation with high doses and by heating the samples. For high enough doses the size of iron aggregates in aluminium becomes so large that eventually superparamagnetism can be established (Fig.7.10). The measurements of the spectra for samples annealed at various temperatures indicated the precipitation of iron clusters and intermetallic compounds occurring at elevated temperatures [7.32]. Various compounds which can be formed at elevated temperatures in Fe-Al system can be identified by their different Mössbauer spectra [7.61]. The annealing procedure is frequently studied for im-

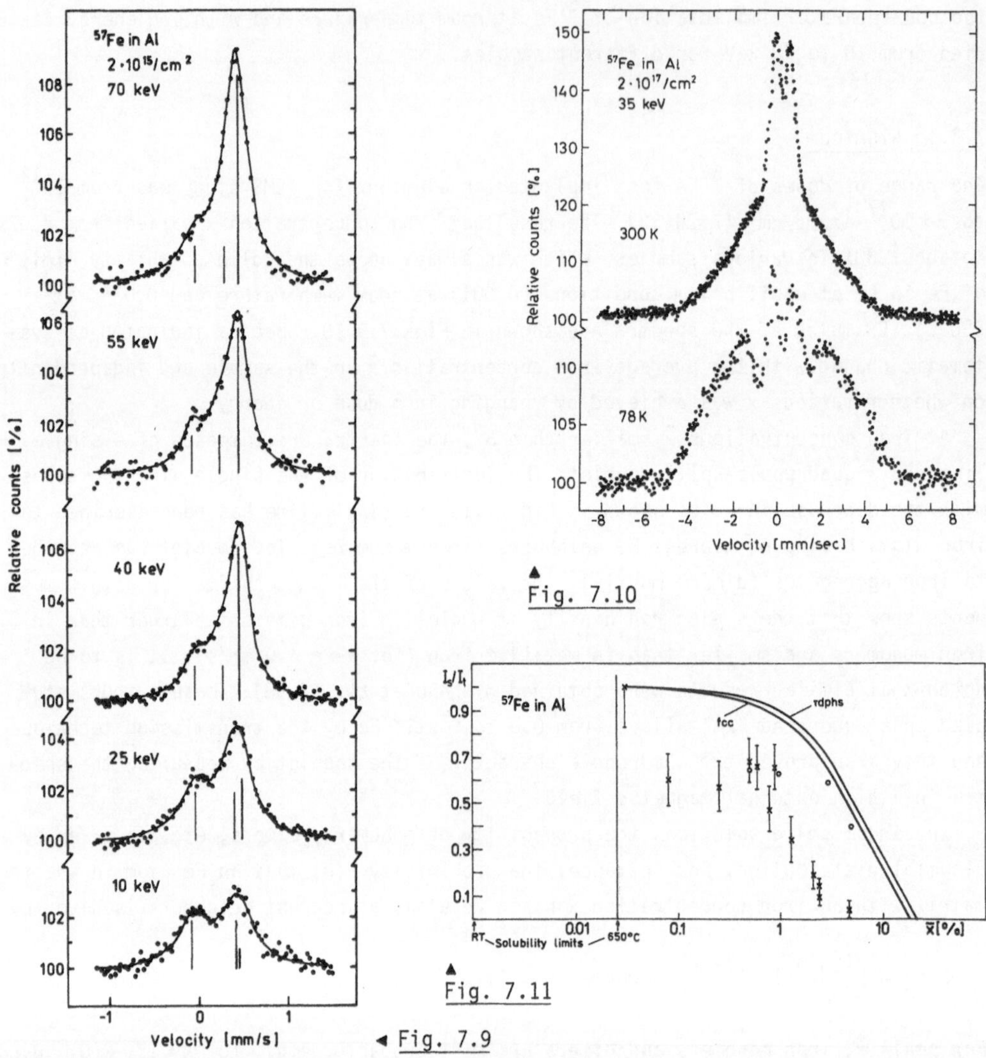


Fig. 7.9. Variation of CEMS spectra of $^{57}$Fe-implanted aluminum with ion energy and, therefore, with concentration of implants ($\bar{x}$=0.5%, 0.6%, 0.9%, 1.1%, and 2.1%, respectively). A single line due to iron monomers and a doublet due to iron dimers are indicated [7.32]

Fig. 7.10. CEMS spectra of $^{57}$Fe-implanted aluminum with a very high dose. The unresolved magnetic hyperfine interaction indicates superparamagnetic aggregates of iron atoms in the sample [7.32]

Fig. 7.11. Intensity of the single-line fraction ($I_s$ relative to the total $I_t$) in the spectra of $^{57}$Fe-implanted aluminum compared to the probability of monomers in random densely packed system of hard spheres (rdphs, upper curve) and in fcc lattice (lower one). Open points present data for alloys obtained by a rapid quenching from melt [7.32]

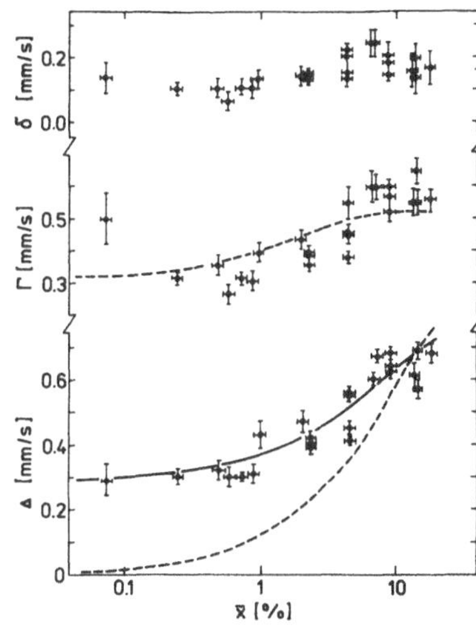


Fig. 7.12. Isomer shift $\delta$, line width $\Gamma$ and quadrupole splitting $\Delta$ of doublet component in CEMS spectra for $^{57}$Fe-implanted aluminum vs concentration of iron. Dashed lines were calculated assuming random distribution of iron atoms. Solid line was obtained assuming increased probability for formation of dimers in calculations of electric field gradients [7.32]

planted samples. Systems obtained by ion implantation are frequently metastable and oversaturated, as in in the case of Fe-Al. The annealing study gives the possibility of observing precipitations of impurity clusters and intermetallic compounds from metastable solid solutions.

Points in Fig.7.11 suggest that splat quenching provided a more random atomic distribution than ion implantation; this is still a question which requires further enlightment. Notice that in spite of the processes of increased iron clustering, iron monomers exist in Al for samples under study at concentrations exceeding highly the solid solubility limits for Fe in Al (Fig.7.11). This is conditioned by a high quenching rate characteristic for both iron implantation and splat quenching techniques.

It was observed that the parameters of the doublet components of the spectra for the "as implanted" samples depend in a systematic way on the average iron concentration $\bar{x}$ in the sample. The quadrupole splitting value $\Delta$ and the line width $\Gamma$ of the doublet component markedly increase with the increase in $\bar{x}$, see Fig.7.12. This effect has been explained on the basis of calculations of electric field gradients in cubic binary alloys [7.32,59,60]. The value of the effective charge defect $q_{eff}(Fe)-q_{eff}(Al) = 0.5$ e, was found. The distributions of the electric field gradients were calculated in a point-charge approximation using a Monte Carlo procedure. The electronic contribution was included by assuming a universal correlation between ionic and electronic contributions to the total electric field gradient. An increased formation of dimers, as well as possible formation of compounds had to be taken into account in order to explain the experimental dependences.

### 7.3.2 Other Metals

Studies of $^{57}$Fe implanted in a number of metals of 3d, 4d, and 5d series were discussed in detail in papers published by the Cracow group [7.3,4,26-28,62-66]. Extensive studies of Cu, Ag, and Au matrices were reported by LONGWORTH and JAIN [7.67-70]. Some results on $^{57}$Fe-implanted 4f and 5f elements were also presented [7.3,4].

In most cases the samples were implanted with $^{57}$Fe of $10^{15}$-$10^{16}$ atoms/cm$^2$ to reach iron concentrations from 0.3% to 3%. The spectra of implanted samples are usually similar to those obtained for dilute iron alloys or $^{57}$Co-diffused sources, and the radiation damage effects were mostly observed only in the line broadening.

Survey of the data revealed [7.62-66] that the relative sizes of iron and host atoms play an important role in a final site characterization, see Fig.7.13. In particular, the spectra of $^{57}$Fe implanted in Fe, Co, Ni, Ru, Rh, Pd, Re, Os, Ir, and Pt matrices (●-points in Fig.7.13) indicated a good substitution of the host atoms by iron implants, and iron dimers were not exhibited by the presence of any additional lines. In V, Cr, Nb, Mo, and Ta (⊗-points) the spectra consisted of a single line and a doublet, which could be ascribed to iron monomers and iron dimers, respectively. The intensities of both components in the spectra indicated a random distribution of iron impurities. The processes of enhanced iron aggregation have so far been observed in matrices such as Al, Cu, Zn, Ag, Cd, and Au (■-points). The phenomena observed for Fe in these hosts are similar to those in rapidly quenched alloys, as was the case for Fe in Al discussed before. It is notable that in bcc lattices a random distribution of iron seems to be favored (V,Cr,Nb,Mo,Ta), while in fcc matrices (Al,Cu,Ag,Au) the processes of aggregation of iron seems to be enhanced. The increased formation of dimers in fcc matrices may be associated with the radiation-enhanced diffusion of iron, possibly via spacious interstitial sites in these matrices.

It should be pointed out that iron aggregation observed in implanted samples is a result of aggregation processes occurring at their very initial stages. The ag-

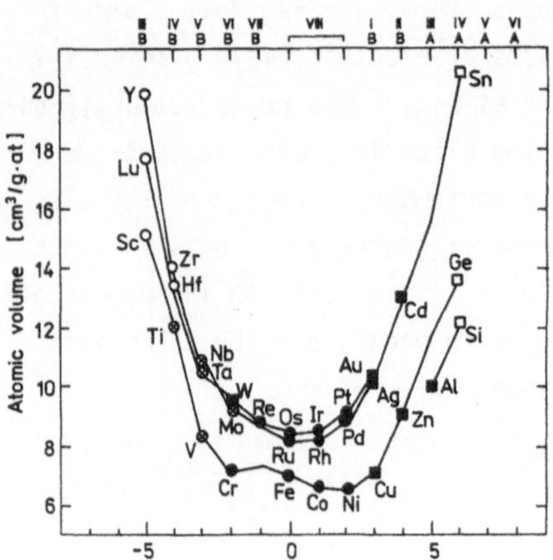


Fig. 7.13. Atomic volumes vs the position of the host element in the Periodic Table of elements with respect to iron. Points are marked differently according to different behavior of iron implants in corresponding matrices (⊗: Fe dimers formed in a random way are observed; ●: Fe dimers not distinguished from Fe monomers; ■: enhanced formation of dimers observed; o: more complex spectra (iron interstitials?); □: p-element hosts, see Sect.7.3.3) [7.66]

gregation may be accelerated at higher temperatures, which can be followed by isochronal thermal annealing procedure. Such studies provide important information on the mechanism of precipitation in alloys. To answer the question whether splat-quenching produces more disordered systems than ion implantation as it appeared in some cases (Al,Cu,Ag) and how it depends on the particular host element as well as on ion dose, temperature during implantation, etc., more data are needed.

In matrices of Sc, Lu, Hf, Y, Zr, and some others, the situation is more complex. These elements are characterized by large atomic volumes and, correspondingly, large interstitial spaces. Additional components observed in Mössbauer spectra for $^{57}$Fe in these hosts are suggested to be associated with iron interstitials but final interpretation of the data has not as yet been given. Here, systematic studies similar to those performed for $^{57}$Fe-implanted Al and Cu, may help to clarify the behavior of iron atoms. Measurements for $^{57}$Fe implanted in metals of 4f and 5f series were also started. All spectra measured until now in rare earths and thorium targets were very similar to those measured in matrices of the last group listed above [7.3,4].

The studies of iron implantation in different host elements may help to establish general rules of alloying, also applicable for systems formed at nonequilibrium conditions. The behavior of iron implants in different metals was recently analyzed [7.66] in terms of the cellur model of alloying, proposed by MIEDEMA et al. [7.71]. This model was originally proposed to describe the formation of alloys and metallic solid solutions under equilibrium conditions. It was found, however, to have a wider applicability. In particular, the Miedema parameters were shown by KAUFMANN et al. [7.72] to describe well the preferences for substitutional and interstitial location of various impurities implanted in beryllium.

In the Miedema model the possibility of alloy formation is determined by the heat of alloy formation $H_f$ which depends on two parameters: the chemical potential $\phi^*$ and the density of electrons at the boundary of the Wigner-Seitz cell $n_{WS}$ for elements. The heat of formation $H_f$ of the alloy depends on the difference in these two parameters for the constituents. The plot in Fig.7.14 shows the alloying possibilities for various elements with iron. Points are designated according to various behavior of iron implants in corresponding matrices. As seen in Fig.7.14 all the matrices discussed here are located in distinct groups in the Miedema diagram for iron alloys. It is also clear that the enhanced formation of iron aggregates in the host elements located in the left quarter of the diagram is associated with positive heat of alloying.

CEMS experiments on $^{57}$Fe-implanted samples were used to determine the s-electron densities and electric field gradients at iron nuclei in monomers and dimers for a number of host elements [7.62,63]. The isomer shift data determined so far are shown in Fig.7.15. The values of $\delta$ for iron monomers are generally consistent with the data for $^{57}$Co-diffused sources, given, e.g., by QAIM [7.73]. The variation of the s-electron density at iron nuclei in monomers depends on the volume available for the impurity atom and on the change in the s-d charge transfer along each transition

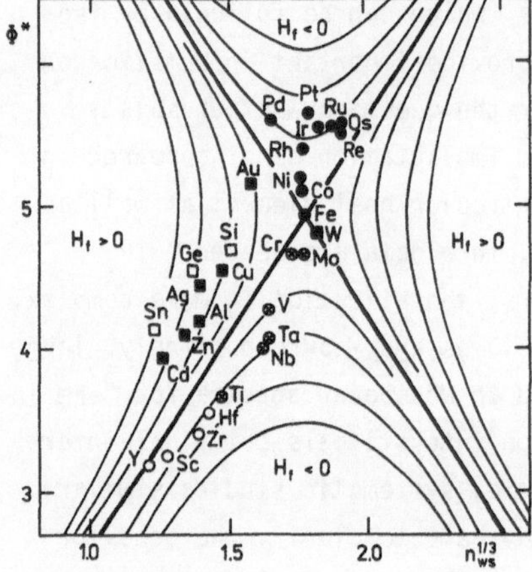


Fig. 7.14. Miedema diagram for the heat of formation of alloys with iron. Straight lines mark $H_f = 0$. Points are designated as in Fig.7.13 [7.66]

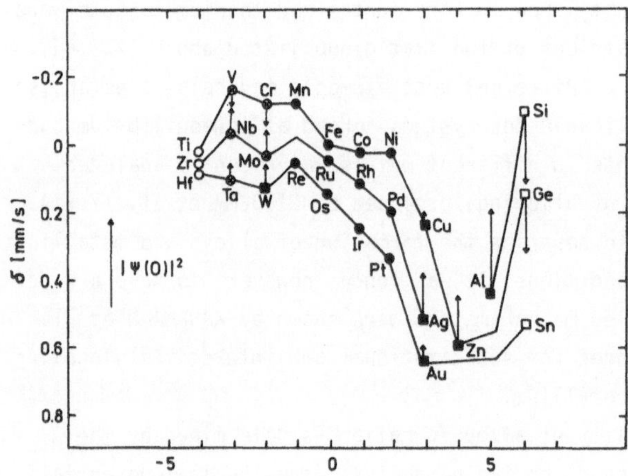


Fig. 7.15. Isomer shifts for Fe monomers and Fe dimers, $\delta_m$ and $\delta_d$, respectively, in various hosts (relative to α-Fe). Points indicate $\delta_m$ and are marked as in Fig.7.13. $\delta_d$-values are indicated by arrows

metal series [7.74,75]. It can be seen that the s-electron density at iron dimers, as determined from CEMS spectra of $^{57}$Fe-implanted samples, is usually closer to the value of s-electron density in pure α-iron and is less affected by the volume effect as compared to iron monomers. Further measurements for other elements are needed to complete the systematics of the data. In particular, data on isomer shifts for Fe in groups of homologous host elements Zn, Cd, Hg and B, Al, Ga, In, Tl are needed for covering the gap between d-elements and diamond-type semiconductors discussed below.

### 7.3.3 Silicon, Germanium and Diamond

Extensive studies of $^{57}$Fe-implanted Si and Ge were made by the Cracow group [7.25, 44,50,76-79]. Samples were implanted with doses between $5 \times 10^{14}$ and $3 \times 10^{16}$ atoms/cm$^2$, which means above the amorphization limits. Iron concentrations $\bar{x}$ were in various samples from 0.1% to more than 10%, exceeding by far the equilibrium solubility (e.g., for Fe in Si the solubility is close to $10^{-4}$% at 1250°C). CEMS spectra of $^{57}$Fe-implanted Si, taken at room temperature after consecutive annealing steps, are shown in Fig.7.16. The spectra indicate lattice reordering processes up to 400°C and precipitation of intermetallic phases α- and β-$FeSi_2$ at high temperatures, as suggested by ZEMČIK and VOJTECHOVSKY [7.80]. Diffusion of iron into the bulk deeper than the electron penetration range is seen at 1200°C.

Mössbauer spectra for Fe implants in (nonheated) Si and Ge, at the whole range of concentrations reported, always presented two broadened lines of (almost) equal intensities. Similar were emission spectra obtained for low concentration of impurities of $10^{14}$ $^{57}$Co atoms/cm$^2$ [7.81]. The spectra were interpreted either as a quadrupole-split doublet [7.25,76] or as two single lines due to iron atoms located at two different, but equally populated positions [7.81].

The quadrupole character of the doublet was finally proven by the CEMS experiment with the magnetic field perturbation technique [7.44]. These measurements were performed at room temperature for the sample of $^{57}$Fe in Si ($\bar{x} \sim 3\%$), using a helium counter described in Sect.7.2 (Fig.7.5) and applying a longitudinal field of a superconducting solenoid. The spectrum measured at 70 kOe (shown in Fig.7.17) is a strong evidence for the quadrupole doublet hypothesis. Evidence for a low-symmetry position of iron implants was therefore obtained. A positive sign and a small asymmetry parameter of the electric field gradient tensor were also inferred from these measurements. Also the Mössbauer emission spectra in an external magnetic field were later taken by LANGOUCHE et al. [7.82-84] for $^{57}$Co implanted in Si and Ge to a dose of $10^{14}$ atoms/cm$^2$ and for Coulomb-excited recoils implanted in Ge host. The quadrupole nature of the spectra was confirmed also in these cases.

It has been pointed out [7.78,79] that the quadrupole splitting and isomer-shift data for $^{57}$Fe-implanted Si and Ge and for amorphous alloys fabricated at high iron concentrations by coevaporation of the components [7.85,86] are consistent and complementary. As seen in Fig.7.18, the quadrupole splitting decreases with an increase in iron concentration, whereas the isomer shift is almost constant in a wide range of $\bar{x}$. A drastic variation in the average electric field gradient at Fe nuclei in Si and Ge is opposite to the change that is observed for Fe in Al. This has been explained by a change in the structure which varies from the random tetrahedral network of covalent bonds in a pure semiconductor to a metallike system of random densely packed hard spheres in an iron-rich phase.

The origin of the electric field gradient at iron nuclei in Si and Ge has been discussed (e.g., [7.4,50]), but no definite site characterization has as yet been proposed. It was suggested [7.50] that implanted Fe atoms come to rest at highly disordered or totally amorphous regions and form there the impurity-vacancy complexes. It cannot be excluded that a very fast diffusion of iron in silicon and germanium enhances some processes of iron aggregation, at concentrations as low as 0.1%. In such a case the quadrupole doublet should be ascribed mainly to small aggregates of iron (dimers?). Such dimers probably occupy interstitial sites in the

Fig. 7.16 ▼ Fig. 7.17 ►

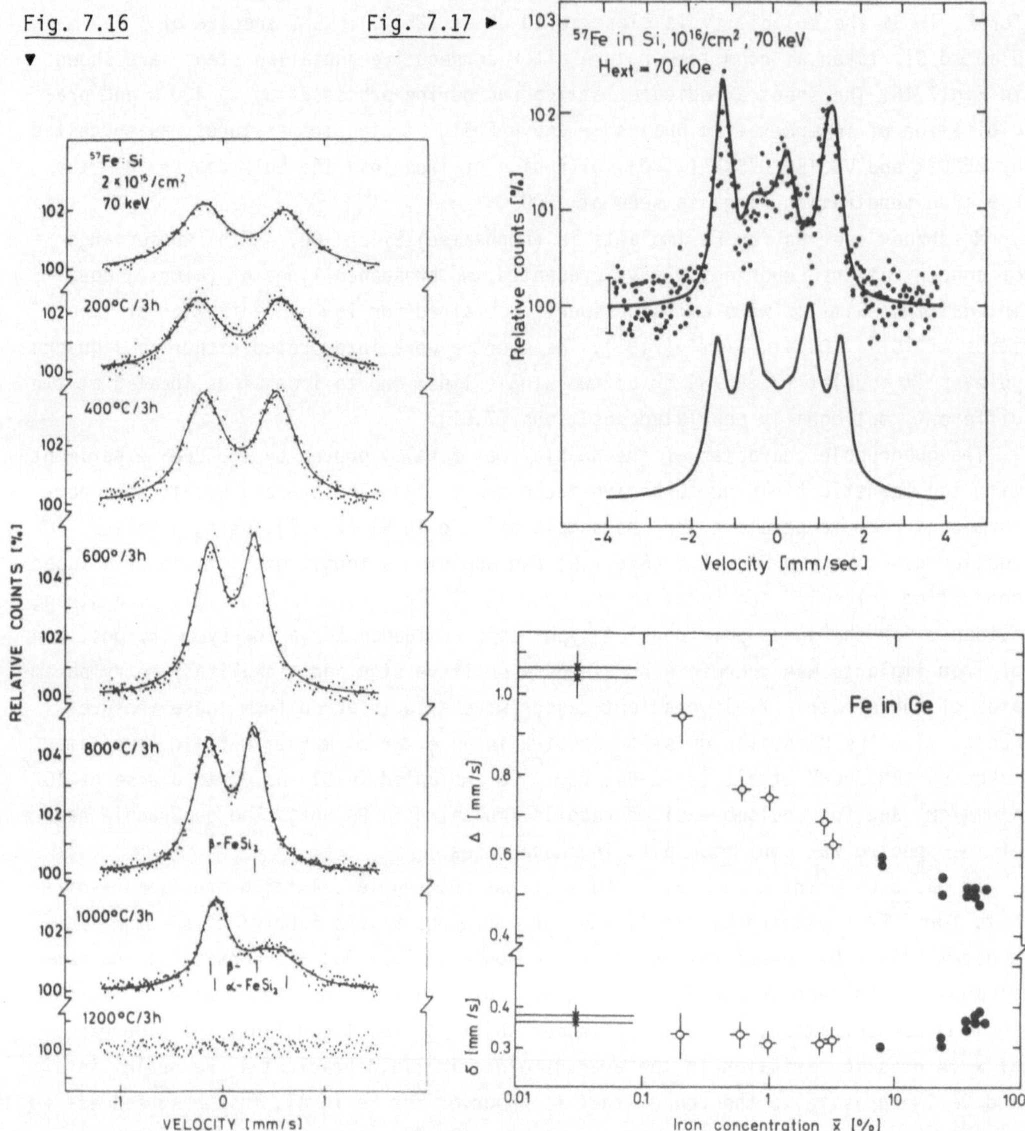


Fig. 7.16. CEMS spectra of $^{57}$Fe-implanted silicon ($\bar{x} \sim 0.5\%$) taken at room temperature after subsequent isochronal heating of the sample. After annealing at 200°C and 400°C lines sharpen, at 600°C $\alpha$-$FeSi_2$ is formed, at 1000°C $\beta$-$FeSi_2$ may also precipitate, at 1200°C iron diffuses into bulk [7.77]

Fig. 7.17. CEMS spectrum for $^{57}$Fe-implanted silicon ($\bar{x} \sim 3\%$) measured in an external longitudinal magnetic field of 70 kOe. The curve presents the spectrum calculated for mixed electric quadrupole and magnetic dipole interactions. The spectrum expected in the case of two independent single lines is shown below for comparison [7.44]

Fig. 7.18. Quadrupole splitting $\Delta$ and isomer shift $\delta$ data for $^{57}$Fe in amorphous FeGe vs iron concentration. (o: data for Fe implants [7.76,78]; x: data for Co implants [7.81]; ●: data for amorphous FeGe films prepared by coevaporation of components [7.85].) [7.78]

tetrahedral network and, when formed, stabilize their amorphous surrounding and stop further clustering. The interpretation presented here is supported by the values of Miedema parameters and positive heat of alloying of Si and Ge with Fe as shown in Fig.7.14. Moreover, it does not contradict the results of recent measurements of the temperature dependence of CEMS spectra [7.50].

Temperature measurements were performed for $^{57}$Fe implanted in both Si and Ge ($10^{16}$ atoms/cm$^2$) [7.50]. Helium counters with cryostat and furnace equipment, as described in Sect.7.2 were used. A rather small temperature dependence of the electric field gradient between 80 K and 500 K was found; at higher temperatures the $\Delta(T)$ dependence was obscured by the irreversible processes in the samples [7.77] (see also Fig.7.16). The character of the $\Delta(T)$ dependence was found to be similar as in the cases when the electric field gradient (EFG) arises from the impurity atom or vacancy located nearby the probe impurity; such $\Delta(T)$ dependences were observed in doped metals, e.g. [7.87]. Similarly small temperature dependence has been found also for the EFG in dimers in $^{57}$Fe in Al [7.32].

Measurements by LANGOUCHE et al. [7.24,88], for $^{57}$Co implanted in Si at low doses, indicate sharp concentration dependence of spectra which vary from an almost symmetric quadrupole doublet at dose of $10^{14}$ atoms/cm$^2$ to nearly a single line at $10^{12}$ atoms/cm$^2$. The single line was attributed to iron atoms located in nondamaged sites in a diamond-type lattice. We are tempted to ascribe a single line to iron monomers located at high-symmetry sites and a doublet to iron dimers in an amorphous network of a semiconductor. With such assignment one can achieve an agreement between all the data available until now. Also the values of the isomer shifts for the two sorts of iron atoms assigned in this way fit well (for both silicon and germanium) to the systematics of the data given in Fig.7.15.

The electron spin resonance (ESR) data of LUDWIG and WOODBURY [7.89] suggest that dilute iron impurities in silicon and germanium exist as interstitials in a charge state of $Fe^0$, and that $Fe^{1+}$ state is present in samples which are quenched rapidly. It can be suggested that the two states observed in implanted samples, and represented by, respectively, the singlet and the doublet in Mössbauer spectra, correspond to the two states observed by ESR; the isomer shifts of the two Mössbauer components do not exclude such assignment. Additional Mössbauer and ESR experiments should be made for establishing full consistency between the data.

Diamond is a fascinating material with many outstanding physical properties. Such properties like a very small atomic volume, high hardness and a very high Debye temperature of diamond ($\theta_D$=2000 K) should influence strongly a state of Mössbauer impurities in $\alpha$-C. Because the solubility limits for various impurities in diamond are very low, the ion implantation is the only promising method of doping (see [7.90,91] for recent references). Implantation behavior of various impurities in diamond is also of special interest because of prospects of diamond-based semiconductor electronics.

Behavior of $^{57}$Co and $^{57}$Fe atoms implanted into diamond was studied by Mössbauer spectroscopy for isotope separator implanted samples [7.21,81] and for Coulomb-excited recoils [7.92]. Recently, the measurements were performed in a wide range of temperatures for the sample implanted with $5\cdot10^{13}$ of $^{57}$Co atoms/cm$^2$ [7.93]. The last measurements revealed the existence of two components in the spectra, a single line and a quadrupole-split doublet, with the single line contribution increasing strongly upon the sample annealing. The single line component indicated also an unusually large s-electron density at iron nuclei ($\delta$=-0.95 mm/s) and a very high effective Debye temperature of iron atoms ($\theta_{eff}\approx$800 K). On this basis the authors concluded that Fe atoms in regular high-symmetry sites (substitutional or tetrahedral interstitial sites) in diamond are exposed to a very high pressure and are highly restricted in their vibrational amplitudes. The last effect offers attractive possibilities for fabricating Mössbauer sources with a very high value of the recoilless fraction.

The low-symmetry iron sites (quadrupole-split doublet) indicate a high electric field gradient and are thought to be connected with damaged regions in the lattice, similarly as in the case of Si and Ge hosts. CEMS spectra of $^{57}$Fe implanted in $\alpha$-C at doses between $10^{14}$ atoms/cm$^2$ and $10^{16}$ atoms/cm$^2$ show a gradual decrease in the quadrupole splitting, from about 2.5 mm/s at low doses to about 1 mm/s at high doses. In analogy to the data for Si and Ge, this change can be attributed to the gradual increase in both the average interatomic distance and the coordination number of iron with rising concentration. Further studies of $^{57}$Fe in diamond may help to elucidate the mechanism of implantation induced diamond-to-amorphous-carbon transition.

It was shown that the s-electron density and electric field gradient at $^{57}$Fe nuclei implanted into diamond-type lattices are correlated in a simple way with the atomic volumes of host matrices, $\alpha$-C, Si, Ge and $\alpha$-Sn [7.79,93]. Two effects are clearly seen (Fig.7.19). First, the s-electron density at the high-symmetry site in a particular matrix is always larger than at the low-symmetry site; the difference for a diamond lattice being extremely high. Moreover, the s-electron density at the Fe nucleus in a high symmetry site in diamond is much larger than in any other elemental host studied so far. Second, the s-electron densities for both sites increase with the decrease in the interatomic distance of the host matrix, with the dependence for the high-symmetry sites being much stronger. Both effects can be explained by a volume compression of Fe atoms at regular sites in matrices under study. For Fe atoms in damaged regions (low-symmetry positions) the lattice compression is greatly reduced. As seen in Fig.7.19, also the value of the electric field gradient at low-symmetry sites increases with the increase in the interatomic distance. It is shown therefore that the volume dependence of hyperfine interactions is present not only in metals but that it is also a strong effect for the IV-group elements.

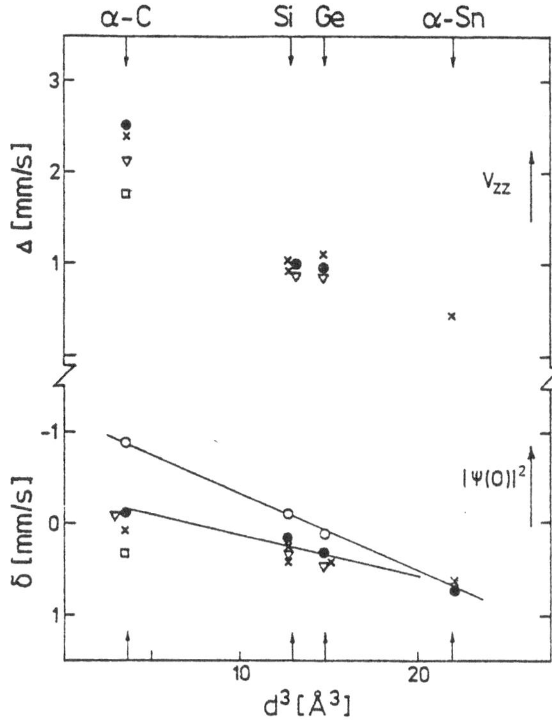


Fig. 7.19. Correlation of the quadrupole splitting Δ and the isomer-shift values δ with the atomic volume in the diamond-type lattices. (●: data for Fe implants [7.76]; □: data for Co implants [7.21]; x: data for Co implants [7.81]; ▽: data for Fe recoils [7.92]; ○: values for the single lines taken from [7.93] for a diamond and from [7.88] for Si and Ge). Arrows indicate the direction of the increase in the s-electron density $|\psi(0)|^2$ and in the electric field gradient $V_{zz}$ at iron nuclei [7.93]

### 7.3.4 Chemical Compounds

CEMS studies of ion implantation in crystals of chemical compounds were initiated recently in Cracow [7.3,94,95] and in Lyon [7.96,97]. The first group investigated $^{57}Fe$ implanted at an energy of 70 keV and a dose of $10^{16}$ atoms/cm$^2$ into a series of alkali halides crystals. The Mössbauer spectra presented well-resolved nearly symmetric quadrupole doublets in LiI, NaCl, KBr, KI and RbCl, unresolved broad patterns in NaF and KF, and more complicated structures in LiF and KCl. In almost all crystals iron was found predominantly in $Fe^{3+}$ state. It was assumed that $Fe^{3+}$ ions are at cationic sites and that in order to compensate for the electric charge the $Fe^{3+}$ ions are associated with two single-charged vacancies, thus forming linear and various planar $V^-$-$Fe^{3+}$-$V^-$ complexes. Calculations of electric field gradients for such complexes were performed, including also dipolar effects associated with a very high polarizability of anions. A principal agreement of the experimental data with the proposed model was obtained. In LiF a substantial fraction of $Fe^{2+}$ ions was also observed, which was attributed to microprecipitates of $FeF_2$ compound. Additionally, metallic iron precipitates can be present, represented by a peak at zero velocity in the spectra. The $FeF_2$ compound was found to be preferentially formed in rapidly quenched iron in LiF solution, as shown by ANAND [7.98].

The group in Lyon investigated LiF and MgO crystals implanted with 100 keV $^{57}Fe$ ions at doses of $6\cdot10^{16}$ atoms/cm$^2$ and $2\cdot10^{16}$ atoms/cm$^2$, respectively [7.96,97]. The results of CEMS experiments were supplemented by optical absorption, transmission electron microscopy and ion channeling data. The creation of $FeF_2$ molecular clusters in LiF was confirmed. The observed increase in the background of the UV region in the optical absorption spectra was interpreted as an indication for (small) metallic iron clusters. In spite of the high number of defects present, the crystal structure checked by the channeling and electron microscopic measurements was found to be preserved even after high dose implantations. (The density of defects in implanted regions was determined to be of the order of $10^{14}$-$10^{5}$/cm$^2$ of F, $F^+$ and V centers in MgO and F centers in LiF.) Thermal annealing of MgO sample at 700°C was found to create iron oxide precipitates and after 800°C ferrite spinel precipitates. The size of the spinel particles was determined, with the mean value of about 20 nm.

A large field of the application and study of radiation effects and hot chemistry reactions induced by ion implantation into insulators is open to further research. In particular, as it is indicated by the examples presented in this section, the role of point defects, the relationship between the relevant equilibrium phase diagrams and formation of compounds at highly nonequilibrium conditions of ion implantation, as well as implantation-induced chemical microsynthesis, can be studied by CEMS technique.

## 7.4 Remarks

The method of conversion electron Mössbauer spectroscopy was successfully applied to study iron implantation. At present, the method is practically applicable for studying $^{57}$Fe doses as low as $5 \cdot 10^{13}$-$10^{14}$ atoms/cm$^2$. Its sensitivity can be still improved, in particular by using the large area multilayer samples and counters and by applying much stronger $^{57}$Co sources than those now in use. The signal-to-noise ratio can be also further reduced, in particular by reducing the photoelectron background in skilfully designed experimental setups. Further extension of the temperature range both for $^{57}$Fe implantation and CEMS measurements is also possible. Application of various other techniques such as ion channeling, laser and electron pulse annealing, transmission electron microscopy, X-ray photoemission, Auger electron spectroscopy, etc. in these studies is also very fruitful. Combination of CEMS technique with implantation of some other Mössbauer isotopes, for instance $^{119}$Sn, $^{151}$Eu, $^{181}$Ta, etc. would be very desirable.

The method presented in this chapter has so far been applied mostly in studies of hyperfine interactions of implanted iron atoms and their aggregation processes in metals, semiconductors and some insulators. The implantation induced lattice damage and subsequent annealing processes are also frequently investigated. Such studies facilitate understanding of the nature of solid phases created under conditions far from thermodynamic equilibrium in various matrices and are closely related to newly developing fields of nonequilibrium micrometallurgy, hot chemistry and physics of implantation-doped semiconductors.

Studies exemplified here are of interest not only for basic research but they are also of great technological importance. Among many possible applications, the problem of radiation damage and swelling phenomena, which occur under high flux neutron irradiation in materials used in nuclear power technology, can be mentioned. Such processes can be simulated and therefore more conveniently studied by ion implantation (because the stopping power of ions is several orders of magnitude higher than for neutrons). Preliminary CEMS results for metallic iron and stainless steel implanted at high doses of $^{57}$Fe were reported [7.65], and more research in this field is expected in the future. Another area of particular technological importance is the ion-beam modification of materials used in the engineering industry for improving the resistance of tools against wear, fatigue, corrosion, etc. [7.99-101]. The durability of metal surfaces can be markedly improved by the implantation of small atoms like nitrogen and carbon. In particular, it was shown that the wear resistance of steel can be greatly improved by the implantation of nitrogen atoms and that the increased resistance is long-lasting and is maintained even when the wear track is several times of the implantation depth [7.102]. CEMS study was very helpful in disclosing the atomistic nature of this interesting and technically important phenomenon [7.102,103]. The formation of nitrides under implantation and the migration of nitrogen atoms under wear to dislocations was found important. The formation of carbides was also observed in CEMS spectra of iron samples implanted with carbon atoms [7.19,104].

The problems of ion implantation and radiation damage will certainly occupy scientists and technologists in the future. The CEMS technique combined with ion implantation appears to be very useful in studying these problems.

## References


7.1 H. de Waard: In *Mössbauer Spectroscopy and Its Applications* (IAEA, Vienna 1972) p.123

7.2 H. de Waard: Phys. Scr. *11*, 157 (1975)

7.3 B.D. Sawicka, J.A. Sawicki: In *Proc. Int. Conf. on Mössbauer Spectroscopy*, Vol.2, ed by D. Barb, D. Tarina (Bucharest 1977) p.35

7.4 B.D. Sawicka: *Hyperfine Interactions of $^{57}$Fe Implanted in Solids Studied by Conversion Electron Mössbauer Spectroscopy*, Rpt No.1030/PL, Institute of Nuclear Physics, Cracow (1978)

7.5 P.D. Townsend, J.C. Kelly, N.E.W. Hartley: *Ion Implantation, Sputtering and Their Applications* (Academic, London, New York, San Francisco 1976)

7.6 F. Eisen, L.T. Chadderton (eds.): *Ion Implantation* (Gordon and Breach, London, New York, Paris 1971)

7.7 G. Dearnaley, J.H. Freeman, R.S. Nelson, J. Stephen: *Ion Implantation* (North Holland, Amsterdam 1973)

7.8 A. Perez, R. Coussement (eds.): *Site Characterization and Aggregation of Implanted Atoms in Materials* (Plenum, New York 1979)

7.9 J. Gyulai (ed.): *Proc. Int. Conf. of Ion Beam Modification of Materials*, Budapest, 1978, Radiat. Eff. (1980)
7.10 "Proc. Int. Conf. of Ion Beam Modification of Materials", Albany, New York 1980, Nucl. Instrum. Methods, in press (1981)
7.11 J. Lindhard, M. Scharff: Phys. Rev. *124*, 128 (1961)
7.12 J. Lindhard, M. Scharff, H.E. Schiøtt: K. Dan. Vidensk. Selsk. Mat. Fys. Med. *33*, No 14 (1963)
7.13 H.E. Schiøtt: Can. J. Phys. *46*, 449 (1968); see also [7.6] p.197 and [7.7] p.766
7.14 R. Behrish (ed.): *Sputtering by Particle Bombardment*, Topics in Applied Physics (Springer, Berlin, Heidelberg, New York 1981)
7.15 U. Gonser: In *Proc. Int. Conf. on Mössbauer Spectroscopy*, Vol.2, ed. by A.Z. Hrynkiewicz, J.A. Sawicki (Cracow 1975) p.113
7.16 G. Vogl: J. Phys. (Paris) Colloq. *35*, C6-165 (1974)
7.17 G. Vogl: Hyperfine Interact. *2*, 151 (1976)
7.18 I. Dézsi: J. Phys. (Paris) Colloq. *41*, C1-11 (1980)
7.19 B.D. Sawicka: In *Nuclear Physics Methods in Materials Research*, ed. by K. Bethge et al. (Vieweg, Braunschweig, Wiesbaden 1980) p.216-234
7.20 G.D. Sprouse, G.M. Kalvius, S.S. Hanna: Phys. Rev. Lett. *18*, 1041 (1965)
7.21 F.S. de Barros, D. Hafemeister, P.J. Viccaro: J. Chem. Phys. *52*, 2865 (1970)
7.22 G. Vogl, W. Mansel, P.H. Dederichs: Phys. Rev. Lett. *36*, 1497 (1976)
7.23 E. Verbiest, H. Pattyn, J. Odeurs: J. Phys. (Paris) Colloq. *41*, C1-431 (1980)
7.24 G. Langouche, I. Dézsi, M. van Rossum, J. de Bruyn, R. Coussement: Phys. Status Solidi *B93*, K107 (1979)
7.25 B.D. Sawicka, J.A. Sawicki, E. Maydell-Ondrusz, S. Lazarski: Phys. Status Solidi *A18*, K85 (1973)
7.26 J.A. Sawicki, B.D. Sawicka, S. Lazarski, E. Maydell-Ondrusz: Phys. Status Solidi *B57*, K143 (1973)
7.27 B.D. Sawicka, J.A. Sawicki: Nukleonika *19*, 811 (1974)
7.28 J.A. Sawicki, J. Stanek, B.D. Sawicka: *Proc. Int. Conf. on Mössbauer Spectroscopy*, Vol.1, ed. by A.Z. Hrynkiewicz, J.A. Sawicki (Cracow 1975) p.209,211
7.29 J. Stanek, J.A. Sawicki, B.D. Sawicka: Nucl. Instrum. Methods *130*, 613 (1975)
7.30 B.D. Sawicka, J.A. Sawicki, J. Stanek: Nukleonika *21*, 949 (1976)
7.31 M.J. Tricker, R.K. Thorpe, J.H. Freeman, G.A. Gard: Phys. Status Solidi *A33*, K97 (1976)
7.32 B.D. Sawicka, M. Drwiega, J.A. Sawicki, J. Stanek: Hyperfine Interact. *5*, 147 (1978)
7.33 F.E. Wagner: J. Phys. (Paris) Colloq. *37*, C6-673 (1976)
7.34 D.C. Champeney: Rep. Prog. Phys. *42*, 1017 (1979)
7.35 K.R. Swanson, J.J. Spijkerman: J. Appl. Phys. *41*, 3155 (1970)
7.36 H. Bokemeyer, K. Wohlfahrt, E. Kankeleit, D. Eckardt: Z. Phys. *A274*, 305 (1975)
7.37 J.J. Spijkerman: In *Mössbauer Effect Methodology*, Vol.7, ed. by I.J Gruvermann Plenum, New York 1971) p.85
7.38 G. Weyer: In *Mössbauer Effect Methodology*, Vol.10, ed. by I.J. Gruvermann (Plenum, New York 1976) p.301
7.39 M.J. Tricker: Surf. Defect Prop. Solids *6*, 106 (1977)
7.40 V.E. Cosselett, R.N. Thomas: Brit. J. Appl. Phys. *15*, 883 (1964)
7.41 R.A. Krakowski, R.B. Müller: Nucl. Instrum. Methods *100*, 93 (1972)
7.42 D. Liljequist, T. Ekdahl, U. Bäverstam: Nucl. Instrum. Methods *155*, 529 (1978)
7.43 S. Damgaard, J.W. Petersen, G. Weyer, O. Holck: Phys. Status Solidi *A58*, 443 (1980)
7.44 B.D. Sawicka, J.A. Sawicki: Phys. Lett. *A64*, 311 (1977)
7.45 Y. Isozumi, M. Takafuchi: Bull. Inst. Chem. Res., Kyoto Univ. *53*, 63 (1975)
7.46 Y. Isozumi, M. Kurakado, R. Katano: Nucl. Instrum. Methods *166*, 407 (1979)
7.47 J.A. Sawicki, B.D. Sawicka, J. Stanek: Nucl. Instrum. Methods *138*, 565 (1976)
7.48 J.A. Sawicki: Nucl. Instrum. Methods *152*, 577 (1978)
7.49 J.A. Sawicki, J. Stanek, B.D. Sawicka, J. Kowalski: Nukleonika *24*, 1161 (1979)
7.50 B.D. Sawicka, J.A. Sawicki, J. Stanek, T. Tyliszczak, J. Kowalski: Phys. Status Solidi *A56*, 451 (1979)
7.51 F. Lecomte, V. Perez-Mendez: IEEE Trans. Nucl. Sci. *NS-25*, 964 (1977)
7.52 O. Massenet: Nucl. Instrum. Methods *153*, 419 (1978)
7.53 T. Tyliszczak, J.A. Sawicki, J. Stanek. B.D. Sawicka: J. Phys. (Paris) Colloq. *41*, C1-117 (1980)

7.54 Zw. Bonchev, A. Jordanov, A. Minkova: Nucl. Instrum. Methods *70*, 36 (1969)
7.55 T. Toriyama, K. Saneyoshi, K. Hisatake: J. Phys. (Paris) Colloq. *40*, C2-14 (1979)
7.56 D. Liljequist, B. Bodlund-Ringström: Nucl. Instrum. Methods *160*, 131 (1979)
7.57 W. Jones, M.J. Tricker, G.A. Gard: J. Mater. Sci. *14*, 751 (1979)
7.58 S. Nasu, U. Gonser, P.H. Shingu, Y. Murakami: J. Phys. F *4*, L24 (1974)
7.59 A. Pustówka, B.D. Sawicka, J.A. Sawicki: Phys. Status Solidi *B57*, 783 (1973)
7.60 B.D. Sawicka, J.A. Sawicki, J. Stanek: Acta Phys. Pol. *A45*, 701 (1974)
7.61 S. Nasu, U. Gonser, R.S. Preston: J. Phys. (Paris) Colloq. *41*, C1-385 (1980)
7.62 B.D. Sawicka, J.A. Sawicki, J. Stanek: Phys. Lett. *A59*, 59 (1976)
7.63 B.D. Sawicka, J.A. Sawicki: J. Phys. (Paris) Colloq. *40*, C2-576 (1979)
7.64 B.D. Sawicka: J. Phys. (Paris) Colloq. *41*, C1-429 (1980)
7.65 B.D. Sawicka: Proc. IBMM-78 Conf. (Budapest), Rad. Effects *48*, 25 (1980)
7.66 B.D. Sawicka: Proc. IBMM-80 Conf. (Albany), Nucl. Instrum. Methods, in press (1981)
7.67 G. Longworth, R. Jain: J. Phys. F *8*, 351 (1978)
7.68 R. Jain, G. Longworth: J. Phys. F *8*, 363 (1978)
7.69 G. Longworth, R. Jain: J. Phys. F *8*, 993 (1978)
7.70 G. Longworth, R. Jain: J. Phys. (Paris) Colloq. *40*, C2-608 (1979)
7.71 A.R. Miedema, R. Boom, F.R. de Boer: J. Less-Common Met. *41*, 283 (1975); *46*, 67 (1976)
7.72 E.N. Kaufmann, R. Vianden, J.R. Chelikovsky, J.C. Phillips: Phys. Rev. Lett. *39*, 1671 (1977)
7.73 S.M. Qaim: Proc. Phys. Soc. (London) *90*, 1065 (1967)
7.74 R. Ingalls: Phys. Rev. *155*, 157 (1967); *B6*, 41 (1972); Solid State Commun. *14*, 11 (1974)
7.75 R.E. Watson, L.H. Bennett: Phys. Rev. *B17*, 3714 (1978)
7.76 B.D. Sawicka, J.A. Sawicki, J. Stanek: J. Phys. (Paris) Colloq. *37*, C6-882 (1976)
7.77 J.A. Sawicki, B.D. Sawicka, J. Stanek, J. Kowalski: Phys. Status Solidi *B77*, K1 (1976)
7.78 J.A. Sawicki, B.D. Sawicka: Phys. Status Solidi *B80*, K41 (1977)
7.79 J.A. Sawicki, B.D. Sawicka: Phys. Status Solidi *B86*, K159 (1978)
7.80 T. Zemčik, K. Vojtéchovsky: Phys. Status Solidi *B66*, K99 (1974)
7.81 G. Weyer, G. Grebe, A. Kettschau, B.I. Deutch, A. Nylandsted-Larsen, O. Holck: J. Phys. (Paris) Colloq. *37*, C6-893 (1976)
7.82 G. Langouche, I. Dézsi, M. van Rossum, J. de Bruyn, R. Coussement: Phys. Status Solidi *B89*, K17 (1978)
7.83 G. Langouche, N.S. Dixon, L. Gettner, S.S. Hanna: J. Phys. (Paris) Colloq. *41*, C1-441 (1980)
7.84 G. Langouche, N.D. Dixon, L.S. Fritz, S.S. Hanna: Hyperfine Interact. *8*, 129 (1980)
7.85 O. Massenet, H. Daver: Solid State Commun. *21*, 25 (1976)
7.86 O. Massenet, H. Daver: Solid State Commun. *25*, 917 (1978)
7.87 A. Weidinger, O. Echt, E. Recknagel, G. Schatz, Th. Wichert: Phys. Lett. *A65*, 247 (1978)
7.88 G. Langouche, I. Dézsi, J. de Bruyn, M. van Rossum, R. Coussement: J. Phys. (Paris) Colloq. *40*, C2-547 (1979)
7.89 G.W. Ludwig, H.H. Woodbury: Phys. Rev. Lett. *5*, 98 (1960)
7.90 R. Kalish, T. Bernstein, B. Shapiro, A. Talmi: Rad. Effects *52*, 153 (1980)
7.91 T. Bernstein, R. Kalish: Proc. IBMM-80 Conf. (Albany); Nucl. Instrum. Methods, in press
7.92 G.L. Latshaw, J.B. Russel, S.S. Hanna: Hyperfine Interact. *8*, 105 (1980)
7.93 B.D. Sawicka, J.A. Sawicki, H. de Waard: to be published
7.94 J. Kowalski, J. Stanek, B.D. Sawicka, J.A. Sawicki, M. Drwiega: J. Phys. (Paris) Colloq. *41*, C1-451 (1980)
7.95 J. Stanek, J. Kowalski, B.D. Sawicka, J.A. Sawicki: Proc. IBMM-80 Conf. (Albany); Nucl. Instrum. Methods, in press (1981)
7.96 A. Perez, J.P. Dupin, O. Massenet, G. Marest, P. Bussière: Rad. Effects *52*, 127 (1980)
7.97 A. Perez, M. Treilleux, L. Fritsch, G. Marest: Proc. IBMM-80 Conf. (Albany); Nucl. Instrum. Methods, in press (1981)

7.98 H.R. Anand: Phys. Status Solidi *B84*, 227 (1977)
7.99 M.J. Tricker: Proc. Conf. on Chemical Applications of Mössbauer Spectroscopy, Houston 1980, in press
7.100 G. Dearnaley: In *Nuclear Physics Methods in Materials Research*, ed. by K. Bethge et al. (Vieweg, Braunschweig, Wiesbaden 1980) p.56-69
7.101 C.M. Breece, J.K. Hirvenen (eds.): *Ion Implantation Metallurgy* (Met. Soc. of AIME, Warrendale 1980)
7.102 G. Dearnaley, N.E.W. Hartley: Thin Solid Films *54*, 215 (1978)
7.103 G. Longworth, N.E.W. Hartley: Thin Solid Films *48*, 95 (1978)
7.104 G. Longworth, R. Atkinson: Proc. Conf. on Chemical Applications of Mössbauer Spectroscopy, Houston 1980, in press

# 8. Selected "Exotic" Applications

R. S. Preston and U. Gonser

In a book with the subtitle "the exotic side of the method" it seems quite appropriate to discuss and explore some odd and strange examples in a separate chapter. Among the wide range of applications found for the Mössbauer effect are a number that are funny or sophisticated or strange or surprising. Such adjectives might be applied to the methods used, the subject investigated, or the results obtained. In this chapter we describe some applications that made a special appeal to our sense of the exotic.

## 8.1 Tests of Relativity

Since the earliest days of Mössbauer spectroscopy, the unprecedented sharpness of the resonance lines has encouraged investigators to carry out tests of special and general relativity that require the detection and measurement of very small frequency shifts. Besides being useful as tests of theory, many of these experiments have proved to be very educational. The controversies that sometimes arose over what was being tested – special relativity, general relativity, the principle of equivalence, Mach's principle – and whether the result was trivial or significant have forced both the investigators and the readers of their results to sharpen their understanding of relativity theory and its observable consequences.

### 8.1.1 Special Relativity

*Relativistic Frequency Shifts*

For atoms vibrating in a solid, the first-order Doppler shift of the Mössbauer lines is, of course, zero. The average of $(v^2/c^2)$, however, is not zero. Although much less than unity, it nevertheless results in a relativistic "thermal shift", which was observed by POUND and REBKA [8.1] at nearly the same time that it was predicted by JOSEPHSON [8.2]. It is interesting to note that the explanation of this effect given by POUND and REBKA was in terms of the "second-order Doppler shift" (also called "the transverse Doppler effect") which is merely the time dilatation. JOSEPHSON's treatment, on the other hand, depended on the relativistic mass difference for the Mössbauer atom in its ground and excited nuclear states. He also mentioned that the predicted magnitude of the effect is identical for the two treatments.

This is a trap for the unwary, since at first sight these appear to be different relativistic effects and therefore additive. JOSEPHSON considered them to be the same effect as viewed from two different standpoints, but a rather detailed calculation of the time-dilatation effect by DEHN [8.3] gave a somewhat different result from JOSEPHSON's. After some discussion in the literature (TROOSTER and BENCZER-KOLLER [8.4], CLARK and STONE [8.5], and DEHN [8.6]) all parties seemed to finally agree that there is only one relativistic shift and that the two methods of calculation give identical results.

A second way to observe the time dilatation without any interference from ordinary first-order Doppler shifts is to fix the Mössbauer source and absorber at different distances from the axis of a system which rotates at a constant angular velocity. Since the distance between the source and the absorber does not vary with time, again there is no first-order Doppler shift. But the mean-squared velocities of the source and absorber are different because they are at different distances from the axis of rotation, and therefore their time dilatations are different. Thus, the frequency of the radiation emitted by the source and of the radiation absorbable by the absorber are both shifted, but by different amounts. The difference in these shifts was first measured by HAY et al. [8.7,8] who kept the source (but not the absorber) on the axis of rotation so that there was no time dilatation (i.e., frequency shift) of the source due to the motion of the rotating system. All of the observed effect was due to the motion of the absorber. The shift was also investigated by CHAMPENEY et al. [8.9-12] who found that, as expected, there was no shift when the source and absorber were fixed diametrically opposite each other at the same distance from the axis of rotation so that $(v^2/c^2)$ was the same at both locations. The results of KÜNDIG [8.13] for his rotor experiments are especially convincing because he obtained detailed Mössbauer spectra with a spectrometer that was actually attached to the rotating system. The expected relativistic shifts are clearly visible in his spectra, for source and absorber at different distances from the axis of rotation.

The time-dilatation interpretations of the vibration and rotation experiments have been difficult for some people to accept because they believe special relativity applies only to inertial reference frames, while a Mössbauer atom that is vibrating or is attached to a rotating system is always being accelerated and therefore cannot be at rest in any inertial frame. The point has been made, however, that the experimenter who makes the measurements and calculates with the Lorentz transformations is at rest in what is a good approximation to an inertial system, namely his laboratory [8.14]. Therefore, he can legitimately calculate the frequency shifts for the source and absorber and the difference between them by using the instantaneous velocities of the two and the formulas of special relativity. Calculations made from the point of view of an observer stationary in the accelerated (rotating or vibrating) frame are discussed in a later section.

Time dilatation has actually been measured with greater precision in mu-meson decay measurements by BAILEY et al. [8.15].

*Nonrelativistic Effects of Acceleration*

There has been some concern that the observed effects are not entirely relativistic, but either wholly or in part straightforward mechanical effects analogous to the effect of vertical acceleration on a pendulum clock. In the rotor experiments, for instance, centripetal forces are applied continuously to the atoms of the source and absorber. This could conceivably produce a change in the separation of the energy levels of a nucleus through, say, a distortion of the electronic cloud about the nucleus and a consequent isomer-shift effect. Other predictions of special relativity, however, have been tested to a much higher degree of precision by non-Mössbauer methods and it has been found unnecessary to make any additional allowances for nonrelativistic effects of the acceleration on the "clocks" that figured in those experiments (NEWMAN et al. [8.16]).

SHERWIN [8.17] and HÖNL and BENNEWITZ [8.18] have calculated upper limits for the effects of acceleration on the rotor and thermal-shift measurements.

*Relativistic Broadening of Mössbauer Lines*

The simplest calculations of the Mössbauer line shape are nonrelativistic and give a Lorentzian line of natural width for atoms in a perfect crystal. The effect of time dilatation on the mean frequency of the resonance has been discussed in the preceding sections. Time dilatation also has an effect on the line *shape* as has been discussed by SNYDER and WICK [8.19], SILSBEE [8.20], KAGAN [8.21], and WEGENER [8.22]. This effect is too small to be observed with $^{57}$Fe, but could produce observable broadening of the lines in $^{67}$Zn experiments and troublesome line broadenings in $^{107}$Ag and $^{109}$Ag experiments done at not very low temperatures. This relativistic second-order broadening should vanish at $T = 0$ K since it is insensitive to the zero-point motions of atoms in a solid.

## 8.1.2 Special Relativity or General Relativity?

*The Time-Dilatation Experiments and General Relativity*

Attempts have been made to interpret the thermal shift (GUPTA and LAL [8.23]) and the results of the rotor experiments (HÖNL and BENNEWITZ [8.18]) as consequences of or demonstrations of general relativity and, in particular, of the equivalence principle. The equivalence principle is a basic feature of general relativity, and it implies the impossibility of distinguishing the local effects of being stationary in a gravitational field from the local effects of being uniformly accelerated in the absence of any gravitational field.

Clearly, Mössbauer atoms that are vibrating in solids or are attached to rotors are in accelerated reference frames, and inertial effects that might be misinterpreted as gravitational would, of course, be detectable by an observer moving with such a system. The gravitational fields that would have equivalent effects in non-accelerated frames are easily calculated or measured. Armed with this information one can calculate the shift between source and absorber resonant frequencies using the gravitational red-shift formula from general relativity. The result is indeed the same as for the time-dilatation calculation.

The equivalence principle, however, does not imply that gravitational and accelerational effects cannot be distinguished at all, but only that is is impossible to do so in experiments carried out over short time periods in confined regions. For instance, it would take a very peculiar distribution of matter to reproduce, by gravitation, the equivalent of the inertial effects of rotation in a nonrotating system. This is because the resultant gravitational field would have to be directed away from a central axis (the rotor axis), increasing in strength with distance from this axis, and capable of simulating Coriolis forces. The fact is that as soon as the restriction to local experiments is removed, gravitational and accelerational effects *are* distinguishable, and it is difficult to justify claims that experiments carried out in accelerated reference frames are tests of gravitational effects.

This point was made more directly by HÖNL and BENNEWITZ [8.18] who analyzed the problem from the point of view of an observer stationed in the rotating system and using general relativity. They found that when they had reached the point in the theoretical development where they could calculate the shift to be expected, they were working with equations that did not depend on the principle of equivalence and implied nothing beyond what is implied by special relativity. Therefore, they said the rotor experiments are tests of special relativity only.

Mach's principle, according to which "the inertial forces (such as centrifugal and Coriolis forces) experienced in an accelerated laboratory are gravitational, having their origin in the distant matter of the universe, accelerated relative to the laboratory," [8.24] does not seem to be an issue here.

### *Ether-Drift Experiments*

The failure of Michelson-and-Morley-type experiments to detect any motion of the earth through space is widely considered to be one of the experimental cornerstones of special relativity. Nevertheless, according to DICKE [8.25], it is possible to believe that special relativity is correct without believing that it absolutely requires a null result for Michelson-Morley experiments. The reason is not that there may be a measurable velocity of the earth relative to empty space or to an ether of space, but rather that in such experiments there might be a detectable effect of the motion of the earth relative to the reference frame defined by the matter in the rest of the universe. As pointed out by DICKE, such a motion is, in principle, detectable through an anisotropy in the distribution of Hubble shifts as measured at the earth. If observed it would not imply any violation of special relativity. In fact, it has been estimated that the solar system actually has a velocity of $\sim 200$ km/s relative to the frame defined by the fixed stars.

On the other hand, the fact that such a relative motion probably does exist and is, in principle, detectable does not guarantee that it would be detectable in a Michelson-Morley experiment. Indeed, if such a motion were ever detected by a Michelson-Morley experiment this would establish a new and interesting feature of nature which would have to be incorporated into general relativity and might have far-reaching consequences for many areas of science. From the point of view of those who already are convinced of the truth of special relativitiy, then any experiment undertaken to detect or place an upper limit on the ether-drift velocity can be reinterpreted as an experiment to detect or place an upper limit on the strength of some unknown mechanism by which inertial motion of an observer relative to the primary reference frame of the fixed stars might cause an anisotropy in the measured velocity of light. Since any such effect ought to be proportional to the velocity u of the measuring apparatus relative to the primary reference frame, the experimentally determined upper limit on the ether-drift velocity can be converted to units of $\gamma u$ where $\gamma$ is the unknown proportionality constant for the postulated effect.

An ammonia-maser ether-drift experiment by CEDARHOLM et al. [8.26] had yielded the most reliable value so far for the upper limit on the velocity of the earth through space. Reinterpreted as a measurement of the effect of motion relative to the fixed stars, this experiment established an upper limit of 3000 cm/s for $\gamma u$.

TURNER and HILL [8.27] then attempted to measure $\gamma u$ in a Mössbauer experiment using a high-speed rotor with the source on the axis. They looked carefully at the frequency shift of the revolving Mössbauer absorber to see if it depended on whether the instantaneous motion of the absorber was in the supposed direction of u or in the opposite direction. Their result for $\gamma u$ was $220 \pm 840$ cm/s. Using 200 km/s as an estimate of u this result implied that $\gamma = (1 \pm 4) \times 10^{-5}$. Clearly this experiment also set a new upper limit for any ether-drift velocity.

CHAMPENEY et al. [8.12] carried out similar measurements which gave for the ether-drift velocity $160 \pm 280$ cm/s. Interpreted in the same way as the TURNER and HILL, this result is equivalent to $\gamma u = 80 \pm 140$ cm/s, which represents a further reduction of the upper limit for $\gamma$ as well as for the possible ether-drift velocity.

### 8.1.3 General Relativity

#### *The Gravitational Red-Shift*

The gravitational red-shift is truly dependent on the equivalence principle. It is one of the three classical, experimentally verifiable consequences of general relativity, the others being the deflection of light in a gravitational field and the advance of the perihelion of Mercury. The detection and measurement of the gravitational red-shift by the Mössbauer effect is described in the chapter by R.V. Pound.

### *Mach's Principle and a Possible Anisotropy of Inertia*

As was mentioned earlier, according to Mach's principle, the inertia of mass as observed locally is a consequence of the presence of the rest of the mass in the observable universe. COCCONI and SALPETER [8.28,29] suggested that if Mach's principle is correct, then any anisotropy in the distribution of matter in our part of the universe might produce an anisotropy in the inertial mass of each individual object in our part of the universe. An anisotropy of inertia would mean that a force of a given magnitude acting on a given body would produce accelerations of different magnitudes if applied in different directions. But the distribution of mass in our part of the universe really *is* anisotropic because of the shape of our galaxy and because of our noncentral location within it. One consequence of this, as calculated by COCCONI and SALPETER, should be a perturbation of the magnetic splitting of nuclear sublevels. The perturbation would depend on the orientation of the effective hyperfine field relative to the axis defined by the line connecting our position to the center of the galaxy.

SHERWIN et al. [8.30] measured Mössbauer spectra of iron samples in applied magnetic fields. The orientations of the magnetic fields relative to the galactic axes were varied, but without any observable effect on the line widths or spacings. From this result the authors concluded that there is no asymmetry of inertial mass greater than $5 \times 10^{-16}$ in our part of the galaxy. Later experiments by other techniques have pushed this limit still lower. It has since been argued by DICKE [8.24] that a rigorous application of Mach's principle leads to the result that if a local anisotropy of inertia does exist, it cannot be detected in local experiments because all matter in the same locality suffers from the same anisotropy.

### *Weyl's Unified Theory of Gravity and Electromagnetism*

A test of a specific theory of general relativity was carried out by HARRIS et al. [8.31]. This was a test of a gauge-invariant geometry, proposed by Weyl in 1918, which was an attempt to combine gravitation and electromagnetism in a unified theory. One of the predictions of Weyl's theory is that the frequencies of spectral lines emitted by atoms can be altered permanently by having them undergo particular interactions with electromagnetic fields. Specifically, if an atom spends a time t at an electrostatic potential V, it will experience permanent shifts in the spacings of its energy levels. These shifts would be proportional to Vt/c, but the proportionality constant is not given by the theory. Or, if an atom moves in a path which encloses magnetic flux, the spacings of its energy levels will again be shifted, this time by an amount proportional to the enclosed flux. The unknown proportionality constant would be the same as before.

Weyl's theory was at first rejected because no such effects had even been observed. All atoms of the same type appeared to have the same energy-level spacings, although their electromagnetic histories could not all have been the same. However, with the discovery of quasars, there was a revival of interest in this theory. The light from quasars exhibits large red-shifts. If these shifts are taken to be Hubble shifts due to the expansion of the universe, they indicate that quasars must be very far away, and the rate at which they radiate energy must then be unaccountably large in order for them to be as bright as they are. On the other hand, if it is assumed that quasars are really much nearer to us, and therefore radiating energy at more believable rates, then it is nearly impossible to account for their large red-shifts. Therefore, it was suggested that anomalously large red-shifts might occur in relatively near astronomical objects if Weyl's theory is correct and if the atoms in these objects have undergone some unusual electromagnetic experiences. Harris and his co-workers tested both of the predictions of Weyl's theory. To test the effect of an electrostatic potential, they made a Mössbauer absorber from a screw which had been part of the high-voltage terminal of a Van de Graaff accelerator for $6 \times 10^{10}$ volt-h. There was no detectable shift between the spectrum of this absorber and the spectrum of an absorber made from a similar screw whose history was more ordinary. To test the effect of motions along paths which enclose magnetic flux, identical Mössbauer absorbers were sent around the world – some eastward and some westward – so that their paths enclosed $\pm 5 \times 10^{17}$ G-cm$^2$. No frequency shifts were observed between the two kinds of absorbers, or between these absorbers and absorbers which had never left the laboratory.

These negative results appear to exclude the Weyl effect as a possible justification for assuming that quasars are relatively near objects with large red-shifts.

## 8.2 Detection and Measurement of Small Motions of Macroscopic Objects

The Mössbauer effect has been used to detect and measure small motions of macroscopic objects at low frequencies.

BONNAZOLA et al. [8.32] have discussed the optimization of the Mössbauer method of measuring amplitudes and frequencies of small acoustical vibrations.

In an especially exotic experiment, BONCHEV et al. [8.33] used the $^{119}$Sn Mössbauer effect to study the breathing motions and normal flickering motions of the abdomens of ants. The rate of air flow to the ants and the temperature of the air were the variables of these experiments. Stannous oxide absorber material was glued to the upper abdomens of several hundred ants which were then placed in containers that were stacked together to form a Mössbauer absorber. The effects of the breathing and flickering motions could be distinguished from those of the crawling motions, and they produced effects on the Mössbauer line shape which were strongly dependent on the temperature and rate of air flow.

It was suggested by HILLMAN et al. [8.34] that the Mössbauer effect be used to observe the vibrational response of various internal parts of the ear to sounds of different frequencies and intensities. Since then, a number of workers have reported on research of this type [8.35-41]. The Mössbauer effect is superior to direct stroboscopic observation for this kind of measurement because very large amplitudes of vibration are required for stroboscopic observations, limiting that method to relatively low frequencies and unnaturally high power levels of the sound. The Mössbauer effect can measure vibrations at lower amplitude (the Ångstrom region) and higher frequencies. Laser interferometry is another alternative, but it is difficult to apply because of the necessity for maintaining a rigidly fixed and reproducible reference plane. In contrast, the Mössbauer method is not seriously affected by infrequent, small disturbances of the reference position. Like the other methods, the Mössbauer method requires that channels be opened in the head and within the ear of the anesthetized or dead subject (animal or human). Thus, the conditions under which the measurements are made are abnormal. For the Mössbauer method the reason for the openings is to permit insertion of the source, which must be attached directly to the moving part, and to permit passage of the gamma radiation to the rest of the spectrometer. Typically, the source and absorber are stationary relative to each other except for the vibrational motion of the source, which occurs when sound waves are incident on the ear of the subject. The counting rate for gamma radiation reaching the detector is recorded with sound-on and sound-off. From the difference between these rates, the displacement amplitude of the source when vibrating is determined. Quantities of interest are the amplitude of the induced vibration relative to the sound intensity and relative to the corresponding amplitude of vibration of other parts of the ear. By using more sophisticated techniques, phase relations can also be measured, and the development in time of the response to an impulse or click can be studied [8.41].

In a historical survey of research on the mechanism of sound analysis in the ear, ZWISLOCKI [8.42] has discussed the considerable impact made by this research on the theory of auditory frequency discrimination.

## 8.3 Modulation of Gamma-Ray Quanta

Any alteration of the spectrum of frequencies emitted by a Mössbauer source may be considered a modulation of the frequency spectrum. Thus, Mössbauer spectroscopy itself may be considered the study of such modulations. Of special interest are modulations of individual gamma-ray quanta that result from changes occurring during the emission or absorption of single quanta. Examples are the diffusional broadening effects treated originally by SINGWI and SJÖLANDER [8.43] and the relaxational effects first discussed by BLUME [8.44]. These effects have developed into important areas of application of the Mössbauer effect in such fields as diffusion, Brownian motion, and magnetism. Some additional exotic results have followed from further investigations of the modulation of single quanta produced by mechanical motions of macroscopic objects and by the interposition of matter into the path of the gamma radiation.

In an early experiment, RUBY and BOLEF [8.45] mounted a Mössbauer source on a quartz crystal that was driven sinusoidally at 20 MHz. The result was a time-varying Doppler shift, producing a sinusoidal frequency modulation (i.e., phase modulation) of the emitted gamma radiation. In the Mössbauer spectrum were a central Lorentzian line and a set of Lorentzian sidebands symmetrically located on either side with a uniform spacing corresponding to the 20 MHz oscillation frequency. The spacing of the sidebands is readily accounted for by standard frequency-modulation theory, but in this and subsequent experiments, the relative intensities of the sidebands have never agreed completely with the simple theory. This is almost certainly due to the lack of a single amplitude of vibration for all the source atoms [8.46-48]. MISHORY and BOLEF [8.48] found good agreement between experimental intensity ratios and the predictions of a theory based on fast relaxation of the ultrasonic phonons injected into a Mössbauer sample by a quartz crystal. Recent experiments [8.49] indicate that if care is taken in coupling the quartz driver to the Mössbauer sample, the experimental line shape comes close to what is predicted for a single amplitude of vibration for all the Mössbauer atoms.

LYNCH et al. [8.50] analyzed the 14-keV radiation that had been emitted from a $^{57}$Fe source and passed through a resonant absorber. Individual quanta of this radiation were detected in delayed coincidence with the 122-keV quanta which preceded them in the sequence of decays from $^{57}$Co to the ground state of $^{57}$Fe. It was found that the time dependence of the probability of detection was no longer exponential and that the details of the measured decay curve depended on the thickness of the absorber and on the amount by which source and absorber resonance frequencies were displaced from each other. The authors found almost complete agreement between their results and calculations based on the consequences of the frequency dependence of the attenuation and phase shifting of radiation passing through an absorber that is at or near resonance. The alteration of the decay curve is directly due to the frequency dependence of the changes in amplitude and phase of the Fourier components originally present in the emitted quantum. Clearly these alterations of amplitude and phase constitute a modulation of the individual quanta. Following this and similar work by WU et al. [8.51] have come a number of experiments involving "time filtering" in which the detector is turned on for a controlled time interval after each formation of the excited Mössbauer level in a source nucleus. If the time interval is chosen properly, a considerable narrowing of the lines of the Mössbauer spectrum can be achieved [8.52,53].

Analogous modulation can occur during near- or on-resonance scattering of Mössbauer radiation. For example, if the source and scatterer are in resonance, the combined effect of the two Lorentzian line shapes can cause the width of the energy distribution of the scattered radiation to be less than the natural width [8.54]. Various combinations of successive interactions of Mössbauer radiation with scatterers and transmitters are possible, and each produces useful information about the materials used. Selective-excitation double-resonance spectroscopy [8.55] involves scattering followed by transmission. Time filtering may also be used with this technique [8.56]. Transmission followed by scattering gives dispersion effects, which appear to be quite sensitive to small energy shifts of either the scattering or the absorption cross section. Many of these effects, including multiple resonant scattering, have been investigated in a systematic way, both theoretically and experimentally, by BARA [8.57].

It has always been clear that if a source is moving with a fixed velocity, all the frequency components of all the quanta emitted in a given direction are shifted by the same amount. GRODZINS et al. [8.58] produced an equivalent shift by instead having all the quanta pass through a nonresonant material that produced a fixed phase shift per unit path length for all frequency components of the recoil-free fraction. This by itself would not have produced any frequency shift, but in addition the material was wedge shaped and was moving at right angles to the ray direction so that the path length in the material was changing uniformly with time. The resulting cumulative bunching (or stretching out) of the phase was the same for all the frequency components of all the quanta traversing the material, and the result was a uniform frequency shift.

CHAMPENEY and WOODHAMS [8.59] observed the broadening of the Mössbauer spectrum of radiation which had been modulated by inhomogeneities in a disk moving at right angles to the gamma-ray direction. CHAMPENEY [8.60] has discussed the application of this to the measurement of small inhomogeneities in solids. ISAAK and PREIKSCHAT [8.61], RUBY et al. [8.62], and HAUSER et al. [8.63] used rotating choppers to produce periodic amplitude and frequency modulation of individual quanta. In each case a more or less well-resolved sideband pattern was observed and, as discussed by VOITOVETSKII and SAZONOV [8.64], all the results are in good agreement with theoretical expectations.

ASHER et al. [8.65] produced sidebands in the spectrum of radiation passing between an ordinary moving source and an ordinary stationary absorber in a standard Mössbauer spectrometer by interposing a resonant absorber that was moving in synchronism with the source and, in addition, was being driven sinusoidally at 7.6 MHz. The authors showed that the sideband structure is in accordance with expectations from classical modulation theory. In order to simplify their theoretical treatment, they found it convenient to consider the frequency spectrum of the radiation arriving at the vibrating resonant absorber as viewed from the rest frame of the absorber. This is a frequency-modulated wave containing sidebands of the sort seen in the Ruby-Bolef experiment and spaced 7.6 MHz apart. As viewed in the rest frame of the vibrating absorber, the central line of this sideband pattern is then filtered out by the single resonance line of the absorber. When this slightly altered frequency spectrum is transformed back to the rest frame of the source, it is seen that it, too, now consists of a central line with sidebands spaced 7.6 MHz apart. The result is more complicated, but still in agreement with theory, when the vibrating absorber has more than one resonance line.

PERLOW [8.66] has performed an experiment that seems to start out in a similar way, but goes further. In this experiment the source was mounted on an oscillating quartz crystal as in the Ruby-Bolef experiment. The source was otherwise stationary, and a completely stationary single-line absorber was interposed between it and the detector. Consequently, the spectrum of the radiation emerging from this absorber was similar to that of the spectrum emerging from the vibrating absorber, as viewed from the rest frame of the vibrating absorber in the experiment of Asher et al. Where the emphasis in the earlier experiment was on how this frequency spectrum transforms to the laboratory system, in this experiment the spectrum in the laboratory system is already known and the emphasis is on the time dependence of the amplitude of the emerging radiation, since an important component of it, the central line of the sideband pattern, has been largely eliminated by the absorber. The author shows quite simply that the result must be a strong amplitude modulation ("quantum beats") of the intensity of the radiation arriving at the detector and that the modulation is synchronized with the vibrations of the quartz crystal. This is confirmed in detail by the experimental results. The method holds promise as a method for measuring small shifts, since the detailed form of the quantum-beat pattern varies significantly with small shifts of the centroid of the absorber cross section relative to the centroid of the source spectrum.

## 8.4 Atmospheric Aerosols

Wherever iron is present one can be sure Mössbauer people are digging for it. Even atmospheric air contains iron, and methods have been found to collect it for Mössbauer studies. Systematic investigations have been performed by Barbara and Michael KOPCEWICZ [8.67,68]. Their Mössbauer absorbers consisted of filters through which approximately 1000 $m^3$ atmospheric air had been pumped. A number of sites in Poland were selected for the sampling. In general, the Mössbauer spectra consisted of a central quadrupole doublet and a six-line magnetic hyperfine pattern. The isomer shift indicated that the atmospheric iron is in the trivalent state. Measurements at various temperatures gave evidence that the aerosol consists of small particles of $\alpha$-$Fe_2O_3$ exhibiting partly superparamagnetic behavior.

A carefully performed analysis has produced a number of significant findings:

1) The amount of iron in the air and its seasonal variation have been evaluated.
2) The contribution of iron to the total air pollution has been determined.

3) The distribution of particle sizes has been measured.
4) The larger iron-containing particles have been found to be, in general, of industrial origin.
5) There are indications that the smaller particles are of extra-terrestrial origin and are partly formed by condensation of meteoric vapor. Information on the stratosphere ↔ troposphere exchange can be obtained.
6) During extensive nuclear explosions performed in the atmosphere in 1963, the variation of the radioactivity in the air coincided with the changes of iron concentration in the atmosphere. There are even indications that in the nuclear bomb tests, $Fe^{57}$ was produced by the nuclear reaction $Fe^{56}(n,\gamma)Fe^{57}$. Thus, the terrestrial natural abundance of $Fe^{57}$ (2.19%) in the atmosphere has increased.

## 8.5 Archaeology and Art

Mössbauer spectra have been used as "fingerprints" in obtaining information on ancient pottery and in fine art. It seems important to mention here that this technique is a nondestructive method of investigation.

### 8.5.1 Pottery

Pottery is of great archaeological interest because it is one of the main sources of information on the artistic and technological skills of our ancestors in various parts of the world. The provocative question, "Is the History of an Ancient Pottery Ware Correlated with it's Mössbauer Spectrum?" was asked by GANGAS et al. [8.69] as the title of their paper. This correlation has been investigated in a number of laboratories using pottery of various ancient cultures [8.69-84]. Review articles have also been published on this topic [8.85,86]. Since the research is aimed at questions of provenance, age, ancient methods, etc., it is necessary to trace physicochemical changes in the pottery, starting from the original raw materials, continuing through the manufacture, and then the long period of exposure to the elements until the present time. Accordingly, the research work has tried to simulate the firing procedures for various clays, as well as their weathering, by thermal cycling. In most cases the spectrum exhibits a magnetic hyperfine pattern with parameters indicating the presence of $\alpha$-$Fe_2O_3$. In addition, a double line often appears in the center, indicating, superparamagnetism. It is believed that the magnetic hyperfine pattern suffers a time-dependent degradation due to erosion. However, it has also been shown that the spectral intensity ratio of the two components depends sensitively on the original firing temperature. Pottery fired in a reducing atmosphere contains a certain amount of $Fe^{2+}$. From the ratio of the intensities of the spectral components associated with $Fe^{3+}$ and $Fe^{2+}$, information regarding the firing conditions can be deduced. Systematic differences in the Mössbauer parameters, which are observed in some cases, are useful in the identification or classification of potsherds. The effect of aging or weathering over several millenia is a crucial problem, depending on environmental conditions, temperature, humidity, grain size, mixture of minerals in the raw materials, etc. Using Mössbauer scattering techniques, the black and red glosses on samples of Greek and Indian ware were investigated nondestructively.

For Egyptian pottery samples an empirical relation has been found that connects the natural radiation dose with the intensity ratio of the two nonmagnetic central peaks, It was suggested that this relation be used for dating ancient pottery.

Mössbauer spectroscopy has been fruitfully applied to the study of pottery of many ancient cultures as summarized below:

| | |
|---|---|
| English | COUSINS et al. [8.70] |
| Greek | GANGAS et al. [8.71]<br>KOSTIKAS et al. [8.72]<br>LONGWORTH et al. [8.77]<br>KOSTIKAS et al. [8.79]<br>LONGWORTH et al. [8.82] |
| Egyptian | EISSA et al. [8.73]<br>EISSA et al. [8.81] |
| Israelian | HESS et al. [8.74] |
| Iranian Turkestan | BOUCHEZ et al. [8.75] |
| French | JANOT et al. [8.76] |
| Islamic | EISSA et al. [8.78] |
| Amazon River | DANON et al. [8.80] |
| Indian | LONGWORTH et al. [8.82] |
| Japanese (pottery and tiles) | MAEDA et al. [8.83]<br>TAKEDA et al. [8.84] |

### 8.5.2 Fine Art

As a National Gallery of Research Project, KEISCH [8.87] described the following iron-bearing pigments and their characteristic spectra: synthetic yellow iron oxides, yellow ochre, raw sienna, burnt sienna, raw umber, red iron oxide (red ochre, mars red, synthetic and natural), black iron oxide (synthetic and natural), brown iron oxides, Van Dyke brown, green earth, Prussian blue, and others.

Yellow iron oxide has the formula FeOOH. It can be produced by chemical precipitation. The precipitation of FeOOH is, in fact, the first stage in the production of magnetic tapes on the basis of $\gamma$-$Fe_2O_3$. Yellow ochre, raw sienna and raw umber are naturally occurring yellow iron-containing earths with varying amounts of silica and alumina. The color is determined mainly by the amount of $\alpha$-FeOOH (Goethite). Red iron-oxide pigments consist of $\alpha$-$Fe_2O_3$ (Hematite). Black iron-oxide is basically $Fe_3O_4$ (magnetite), an inverse spinel. Mössbauer spectroscopy allows one to determine impurities, site occupation, and the ratio of $Fe^{3+}$ to $Fe^{2+}$. All of these pigments as well as brown iron oxide, a mixture of oxides, exhibit magnetic hyperfine patterns at room temperature. However, due to the fine grain size, which also affects the color, superparamagnetic components are superimposed.

Van Dyke brown is a peatlike material containing only about 1% iron. It shows no indication of magnetic splitting – at least not at room temperature. Green earth is a clay mineral with a complex layered silicate structure. Hydrous iron is in the two-valent state.

Finally Prussian blue is a synthetic pigment with the chemical formula $Fe_4(Fe[CN]_6)$, first synthesized in 1704. It is believed that light-induced fading may occur in Prussian blue. Experiments are underway in a "fadeometer" using spectrophotometry and Mössbauer spectroscopy to investigate the changes associated with the fading.

In these studies by KEISCH [8.87], transmission as well as scattering techniques have been applied. For the latter, a special apparatus was designed and constructed for field investigation. There are two significant goals of this research:

1) Providing assistance in the attribution of works of art to particular artists or schools.
2) Aiding in the identification of fakes and forgeries.

One day Mössbauer spectra may be used for these purposes and perhaps play a role in the law courts.

## 8.6 Medicine and Biology

Fracture, fatigue, recovery, and aging are among the expressions commonly used in describing the macroscopic phenomena of physical metallurgy, and they are also used in medicine. To this list one might even add warts and the pest (or plague) – as the white → gray transition of tin has been called. It is said that this transformation contributed to Napoleon's defeat in Russia when the buttons on the French uniforms got this "disease" at low temperatures. In fact, it is recognized by museum curators that the disease is actually contagious [8.88].

In recent decades we have learned to understand the phenomena and macroscopic ailments of solids on an atomistic scale, i.e., in terms of defects like dislocations, grain boundaries, vacancies, impurities, etc. Microscopic techniques have contributed significantly to this, and it was especially kind of Nature to provide us with $Fe^{57}$ – not only the best isotope in Mössbauer spectroscopy, but also an isotope of the most important element in physical metallurgy.

In contrast, the situation in medicine and biology is much more complex. Many intertwined processes exist. Nevertheless, in further pursuit of the analogy, one might ask, "Is it possible, using microscopic methods like Mössbauer spectroscopy, to find effects that are related to biological defects or, in other words, to real diseases?" In searching for such effects, one is immediately confronted by great disadvantages and difficulties. Unfortunately, carbon, nitrogen, and oxygen do not have appropriate nuclear excited states. One can hardly imagine the extent of the activities that would be going on now in most laboratories for biological research if a carbon isotope existed that was suitable for Mössbauer spectroscopy.

Nevertheless, the fact that iron is present in biological molecules was sufficient for the potential of the method to be tested. Since the first "Mössbauer Effect in Blood" [8.89] in 1963 more than 500 publications have appeared. Detailed information regarding electronic states and ligand bonding, particularly of heme proteins, is now available [8.90-93]. Recently, the molecular motion in myoglobin was investigated by Rayleigh scattering of Mössbauer radiation [8.94].

In using Mössbauer spectroscopy as a diagnostic tool in medicine, we are faced with the problem of the low iron content of biological tissues. Enrichment of $Fe^{57}$ is difficult and costly. In spite of this, $Fe^{57}$ has been used successfully in some cases as a probe to obtain information concerning diseases or defects.

JOHNSON [8.95] reported on dried pneumoconiotic material from human lungs. For a lung infected by hemosiderosis, a fatal disease contracted by coal miners, the Mössbauer absorption is larger than normal by an order of magnitude. In addition, hyperfine splitting occurs at low temperatures, indicating that the absorbing material, a finely divided compound, has a low molecular weight.

Although the amino-acid sequences vary for different hemoglobins, they often have closely similar conformations, similar electronic structures of their heme irons, and similar functional behavior. Amino-acid replacement or exchange can lead to pathological hemoglobins which cause specific abnormalities. In Zürich lives a person with hemoglobin in which the normal distal-histidine sites on the β-chain are occupied by arginine. This is called Hemoglobin-Zürich or Hb $Z H\beta^{63 his \rightarrow arg}$. Its unusual spectral properties were studied in detail by WINTERHALTER et al. [8.96]. At 80 K ordinary deoxy-Hb exhibits a unique quadrupole splitting ($\Delta E_Q = 2.2$ mm/s) and isomer shift ($\delta = 0.9$ mm/s) relative to α-Fe which are typical of high-spin $Fe^{2+}$. In contrast, the spectrum of deoxy-Hb-ZH is a superposition of two components. One corresponds to the iron in the α-chain and has the parameters of ordinary deoxy-Hb. The other component corresponds to the iron in the β-chain and has a small quadrupole splitting ($\Delta E_Q = 0.3$ mm/s) and an isomer shift ($\delta = 0.24$ mm/s) typical of low-spin $Fe^{2+}$. Thus, the electronic structure of deoxy-Hb ZH is significantly different from the normal deoxy-Hb.

BAUMINGER et al. [8.97] investigated red blood cells from patients with β-thalassemia, sickle-cell anemia, and unstable Hb Hammersmith. In all cases a third spectral component was superimposed on those of the two normal constituents, oxy-Hb and deoxy-Hb. The origin of the third component could be identified as iron stored in ferritin or hemosiderin. It was argued that the large ferritin fraction may be caused by the high rate of intracellular denaturation of Hb during the accelerated erythropoieses which take place in these diseases.

Finally a recent discovery on bacteria should be mentioned. The ability of bacteria to orientate themselves has always been a puzzle. FRANKEL et al. [8.98] studied magnetic bacteria that swim in a preferred direction relative to the earth's magnetic field. It was found that some of them contain intracytoplasmic "crystallites" about 500 Å in diameter. Mössbauer spectroscopy revealed that the crystallites consist of $Fe_3O_4$ (magnetite). These arrange themselves in chains of 22 to 25 crystallites that can align in the geomagnetic field. The chains can be regarded as internal compasses for bacterial navigation. Bacteria from aquatic environments in New Zealand and Australia orient in the earth's magnetic field to the South while bacteria in the Northern Hemisphere have the reverse polarity.

## References


8.1 R.V. Pound, G.A. Rebka, Jr.: Phys. Rev. Lett. *4*, 337 (1960)
8.2 B.D. Josephson: Phys. Rev. Lett. *4*, 341 (1960)
8.3 J.T. Dehn: Phys. Lett. *29A*, 132 (1969)
8.4 J. Trooster, N. Benczer-Koller: Phy . Lett. *30A*, 27 (1969)
8.5 M.G. Clark, A.J. Stone: Phys. Lett. *30A*, 144 (1969)
8.6 J.T. Dehn: Phys. Lett. *32A*, 239 (1970)
8.7 H.J. Hay, J.P. Schiffer, T.E. Cranshaw, P.A. Egelstaff: Phys. Rev. Lett. *4*, 165 (1960)
8.8 T.E. Cranshaw, H.J. Hay: *Proceedings of the International School "Enrico Fermi"*, Varenna, 1961 (Academic Press, New York 1961) p.220
8.9 D.C. Champeney, P.B. Moon: Proc. Phys. Soc. (London) *77*, 350 (1961)
8.10 D.C. Champeney, G.R. Isaak, A.M. Khan: Nature *198*, 1186 (1963)
8.11 D.C. Champeney, G.R. Isaak, A.M. Khan: Phys. Lett. *7*, 241 (1963)
8.12 D.C. Champeney, G.R. Isaak, A.M. Khan: Proc. Phys. Soc. (London) *85*, 583 (1965)
8.13 W. Kündig: Phys. Rev. *129*, 2371 (1963)
8.14 N. Sama: Am. J. Phys. *37*, 832 (1969)
8.15 J. Bailey, K. Borer, F. Combley, H. Drumm, F. Krienen, F. Lange, E. Picasso, W. von Ruden, F.J. Farley, J.H. Field, W. Flegel, P.M. Hattersley: Nature *268*, 301 (1977)
8.16 D. Newman, G.W. Ford, A. Rich, E. Sweetman: Phys. Rev. Lett. *40*, 1355 (1978)
8.17 C.W. Sherwin: Phys. Rev. *120*, 17 (1960)
8.18 H. Hönl, F. Bennewitz: Z. Naturforsch. *21A*, 867 (1966)
8.19 H.S. Snyder, G.C. Wick: Phys. Rev. *120*, 128 (1960)
8.20 R.H. Silsbee: Phys. Rev. *128*, 1726 (1962)
8.21 Yu. Kagan: Sov. Phys.-JETP *20*, 243 (1965)
8.22 H.H.F. Wegener: Z. Phys. *A281*, 183 (1977)
8.23 G.P. Gupta, K.C. Lal: Phys. Lett. *36A*, 421 (1971)
8.24 R.H. Dicke: Phys. Rev. Lett. *7*, 359 (1961)
8.25 R.H. Dicke: *The Theoretical Significance of Experimental Relativity* (Gordon and Breach, New York, London 1965)
8.26 J.P. Cedarholm, G.F. Bland, B.L. Havens, C.H. Townes: Phys. Rev. Lett. *1*, 342 (1958)
8.27 K.C. Turner, H.A. Hill: Phys. Rev. *134*, B252 (1964)
8.28 G. Cocconi, E. Salpeter: Nuovo Cimento *10*, 646 (1958)
8.29 G. Cocconi, E.E. Salpeter: Phys. Rev. Lett. *4*, 176 (1960)
8.30 C.W. Sherwin, H. Frauenfelder, E.L. Garwin, E. Lüscher, S. Margulies, R.N. Peacock: Phys. Rev. Lett. *4*, 399 (1960)
8.31 E.G. Harris, P.G. Huray, F.E. Obenshain, J.O. Thomson, R.A. Villecco: Phys. Rev. *7D*, 2326 (1973)
8.32 G.C. Bonazzola, E. Chiavassa, V. Maxia: Nucl. Instrum. Methods *84*, 257 (1970)

8.33 T. Bonchev, I. Vassilev, T. Sapundzhiev, M. Evtimov: Nature *217*, 96 (1968)
8.34 P. Hillman, H. Shechter, M. Rubinstein, S. Shtrikman: Rev. Mod. Phys. *36*, 360 (1964)
8.35 P. Gilad, S. Shtrikman, P. Hillman, M. Rubinstein, A. Eviatar: J. Acoust. Soc. Am. *41*, 1232 (1967)
8.36 B.M. Johnstone, A.J.F. Boyle: Science *158*, 389 (1967)
8.37 B.M. Johnstone, K.J. Taylor, A.J. Boyle: J. Acoust. Soc. Am. *47*, 504 (1970)
8.38 W.S. Rhode: J. Acoust. Soc. Am. *49*, 1218 (1971)
8.39 B.M. Johnstone, G.K. Yates: J. Acoust. Soc. Am. *55*, 584 (1974)
8.40 W.S. Rhode, L. Robles: J. Acoust. Soc. Am. *55*, 588 (1974)
8.41 W.S. Rhode: In *Basic Mechanisms in Hearing*, ed. by A. Moller, P. Boston (Academic Press, New York 1975) pp.49-67
8.42 J.J. Zwislocki: American Scientist *69*, 184 (1981)
8.43 K.S. Singwi, A. Sjölander: Phys. Rev. *120*, 1093 (1960)
8.44 M. Blume: Phys. Rev. Lett. *14*, 96 (1965)
8.45 S.L. Ruby, D.I. Bolef: Phys. Rev. Lett. *5*, 5 (1960)
8.46 T.E. Cranshaw, P. Reivari: Proc. Phys. Soc. (London) *90*, 1059 (1967)
8.47 A. Abragam: *L'Effet Mössbauer* (Gordon and Breach, New York, London 1964) pp.22-24
8.48 J. Mishory, D.I. Bolef: In *Mössbauer Effect Methodology*, Vol.4, ed. by I.J. Gruverman (Plenum Press, New York 1968) pp.13-35
8.49 A.R. Mkrtchyan, G.A. Arutyunyan, A.R. Arakelyan, R.G. Gabrielyan: Phys. Status Solidi *B92*, 23 (1979)
8.50 F.J. Lynch, R.E. Holland, M. Hamermesh: Phys. Rev. *120*, 513 (1960)
8.51 C.S. Wu, Y.K. Lee, N. Benczer-Koller, P. Simms: Phys. Rev. Lett. *5*, 432 (1960)
8.52 D.W. Hammill, G.R. Hoy: Phys. Rev. Lett. *21*, 724 (1968)
8.53 P. Thieberger, J.A. Moragues, A.W. Sunyar: Phys. Rev. *171*, 425 (1968)
8.54 K.P. Mitrofanov, V.P. Gor'kov, M.V. Plotnikova: Nucl. Instrum. Methods *144*, 263 (1977)
8.55 A.N. Artem'ev, G.V. Smirnov, E.P. Sepanov: Sov. Phys.-JETP *27*, 547 (1963)
8.56 B. Balko, G.R. Hoy: Phys. Rev. *B10*, 36 (1974)
8.57 J.J. Bara: *Investigation of Double Mössbauer Resonances* (Institute of Physics, Cracow 1978)
8.58 L. Grodzins, E.A. Phillips: Phys. Rev. *124*, 774 (1961)
8.59 D.C. Champeney, F.W.D. Woodhams: Phys. Lett. *20*, 275 (1966)
8.60 D.C. Champeney: Br. J. Appl. Phys. *18*, 549 (1967)
8.61 G.R. Isaak, E. Preikschat: Phys. Lett. *38A*, 257 (1972)
8.62 S.L. Ruby, R.S. Preston, C.E. Skov, B.J. Zabransky: Phys. Rev. *A8*, 59 (1973)
8.63 U. Hauser, W. Neuwirth, N. Thesen: Phys. Lett. *49A*, 57 (1974)
8.64 V.K. Voitovetskii, S.B. Sazonov: Phys. Lett. *61A*, 261 (1977)
8.65 J. Asher, T.E. Cranshaw, D.A. O'Connor: J. Phys. *A7*, 410 (1974)
8.66 G.J. Perlow: Phys. Rev. Lett. *40*, 896 (1978)
8.67 B. Dzienis (Kopcewicz), M. Kopcewicz: Tellus *25*, 213 (1973)
8.68 B. Kopcewicz, M. Kopcewicz: Tellus *30*, 562 (1978)
8.69 N.H.J. Gangas, I. Sigalas, A. Moukarika: J. Phys. (Paris) *C6*, 867 (1976)
8.70 D.R. Cousins, K.G. Dharmawardena: Nature (London) *223*, 733 (1969)
8.71 N.H. Gangas, A. Kostikas, A. Simopoulos, J. Vocotopoulou: Nature (London) *229*, 485 (1971)
8.72 A. Kostikas, A. Simopoulos, N.H. Gangas: J. Phys. (Paris) *C1*, 107 (1974)
8.73 N.A. Eissa, H.A. Sallam, L. Keszthelyi: J. Phys. (Paris) *C6*, 569 (1974)
8.74 J. Hess, I. Perlman: Archaeometry *16*, 137 (1974)
8.75 R. Bouchez, J.M.D. Coey, R. Coussement, K.P. Schmidt, M. van Rossum, J. Aprahamian, J. Deshayes: J. Phys. (Paris) *C6*, 541 (1974)
8.76 Ch. Janot, P. Delcroix: J. Phys. (Paris) *C6*, 557 (1974)
8.77 G. Longworth, S.E. Warren: Nature (London) *255*, 625 (1975)
8.78 N.A. Eissa, H.A. Sallam, S.M. Negm: J. Phys. (Paris) *C6*, 873 (1976)
8.79 A. Kostikas, A. Simopoulos, N.H. Gangas: J. Phys. (Paris) *C6*, 537 (1974)
8.80 J. Danon, C.R. Enriquez, E. Mittievich, M. Coutinho Beltrão: J. Phys. (Paris) *C6*, 866 (1976)
8.81 N.A. Eissa, H.A. Sallam, M.H. Morcy: J. Phys. (Paris) *C2*, 462 (1979)
8.82 G. Longworth, M.S. Tite: J. Phys. (Paris) *C2*, 460 (1979)
8.83 Y. Maeda, H. Sakai, S. Onoyama, E. Yoshida: J. Phys. (Paris) *C2*, 485 (1979)

8.84 M. Takeda, O. Kawakami, H. Kobayashi, T. Tominaga: J. Phys. (Paris) *C2*, 483 (1979)
8.85 J.M.D. Coey: In *Proc. Int. Conf. Mössbauer Spectroscopy*, Vol.2 (Cracow, Poland 1975) p.333
8.86 A. Kostikas, A. Simopoulos, N.H. Gangas: *Applications of Mössbauer Spectroscopy I*, ed. by R.L. Cohen (Academic Press, New York 1975)
8.87 B. Keisch: J. Phys. (Paris) *C6*, 151 (1974)
8.88 C.L. Mantell: *Tin* (Rheinhold, New York 1949)
8.89 U. Gonser, R.W. Grant, J. Kregzde: Appl. Phys. Lett. *3*, 189 (1963); *4*, 23 (1964)
8.90 G. Lang: Q. Rev. Biophys. *3*, 1 (1970)
8.91 A. Trautwein: Struct. Bonding *20*, 101 (1974)
8.92 P.G. Debrunner; W.T. Osterhuis and K. Spartalian; G. Lang: *Application of Mössbauer Spectroscopy*, Vol.1, ed. by R.L. Cohen (Academic Press, New York, San Francisco, London 1976)
8.93 Y. Maeda: J. Phys. (Paris) *C2*, 514 (1979)
8.94 Yu.F. Krupyanskii, F. Parak, E.E. Gaubmann, F.M. Wagner, V.I. Goldanskii, R.L. Mössbauer, I.P. Suzdalev, F.J. Litterst, H. Vogel: J. Phys. (Paris) *C1*, 489 (1980)
8.95 C.E. Johnson: Phys. Today *24*, 35 (1971)
8.96 K.H. Winterhalter, E.E. DiIorio, J.B. Beetlestone, J.B. Kushimo, H. Uebelhack, H. Eicher, A. Mayer: J. Mol. Biol. *70*, 665 (1972)
8.97 E.R. Bauminger, S.G. Cohen, S. Ofer, E.A. Rachmilewitz: J. Phys. (Paris) *C2*, 502 (1979)
8.98 R.B. Frankel, R.P. Blakemore: J. Magn. Magn. Mater. *15-18*, 1562 (1980)

# 9. The Discovery of the Magnetic Hyperfine Interaction in the Mössbauer Effect of $^{57}$Fe

S. S. Hanna

With 2 Figures

It was the autumn of 1959 and the Iron Age, as reckoned on the Mössbauer calendar [9.1], was just dawning. The phenomenon of recoilless gamma-ray absorption, discovered by Rudolph MÖSSBAUER [9.2] in Ir, had been confirmed the previous summer [9.3, 4], and we had heard via the grapevine [9.5] that the effect was indeed much larger and more sensitive in the nucleus $^{57}$Fe. These new developments were a favorite topic at the lunch tables of the Argonne National Laboratory. What could be done with this sensational new effect? Were we on the threshold of exciting new discoveries? One of the intriguing ideas that came up was the possibility of measuring the shift due to the centripetal acceleration in a high-speed rotor or centrifuge. One day after lunch six of us – Juergen Heberle, Carol Littlejohn, Gil Perlow, Dick Preston, Dieter Vincent and myself – got together and decided to give it a try. We received the immediate and enthusiastic support of our director Mort Hamermesh (who also made invaluable contributions to our investigations) and of the Argonne Laboratory. A special "red shift" account was set up and all the needed facilities of the Laboratory were put at our disposal.

Within a very short time $^{57}$Co activity had been produced on the Argonne cyclotron, and Gil and Mina Rea Perlow were busy trying to make suitable sources of $^{57}$Co diffused into iron metal. Frank Karasek went to work rolling very thin iron foils, and each day he produced a new record for thinness. The detector group fabricated very thin NaI crystals to detect the low-energy $\gamma$ rays, and we were soon ready to try to see the effect in $^{57}$Fe. We needed a suitable "drive", but felt we did not have time to design and construct one. It was suggested that a lathe could be used, so we went to the foreman of our machine shop and asked for his best lathe. At first, he was reluctant to consign his best instrument to a bunch of physicists, but when we mentioned the words "red shift", he was happy to comply. We set up the experiment using the drive mechanism of the lathe carriage to produce the needed slow, smooth motion. We worked at night so as to eliminate vibrations from all the other machinery in the shop. As I recall we were successful in seeing a Mössbauer "dip" in our first full-fledged attempt.

Meanwhile, we were busy fabricating and testing small air-jet-driven rotors made of aluminum. With the help of many people we soon produced a successful model. We ran the experiment several times and undoubtedly observed the "red shift", but we were never satisfied that we had reduced the effect of vibration enough to report the experiment.

However, during these attempts it became clear to us that much more interesting physics was waiting in the wings. In an effort to increase the size and sensitivity of the effect we tried several different chemical forms of source and absorber. We observed the intriguing result that only with identical source and absorber (in our case metal) did we observe maximum absorption at zero velocity. We also reported, "There is evidence that the width of the transmission dip is influenced by the environment of the emitter or the absorber" [9.6].

We did not perceive clearly at the time that these observations predicted the presence of hyperfine interactions. Instead, we followed another provocative line of thought. Very little was known to us then about internal fields in magnetic media. But it was argued that in iron the nucleus was probably in some kind of magnetic field and the nuclear transitions should be polarized according to the "classical Zeeman effect". A simple experiment was devised to test this idea. We obtained some "five-and-dime" magnets and by construction of a simple apparatus [9.7] arranged to measure the Mössbauer absorption when the direction of magnetization in the absorber was set at different angles relative to that in the source. We reasoned that the absorption should be greatest when the magnetizations (i.e., polarizations) were aligned and least when they were perpendicular. One evening we were ready to try this experiment and were rewarded by a beautiful sine curve for the Mössbauer ab-

Fig. 9.1. The lathe arrangement used for measuring the hyperfine spectrum of $^{57}Fe$. From right to left: the large magnet used to align the source, the small permanent magnet and apparatus used for rotating the alignment of the absorber, and the NaI detector. Photograph by David Linton previously published in Scientific American. Reproduced here by kind permission of Mr. Linton and Scientific American

sorption as the angle between polarizations was varied from $0^{\circ} \rightarrow 180^{\circ}$ [9.7]. This occurred a few days before the holiday season in December and contributed much to the festive spirit.

We were now convinced that some form of magnetic hyperfine interaction was present at $^{57}$Fe in iron metal. We redoubled our efforts to explore the splitting in the Mössbauer spectrum of metallic iron. We knew that only four "hyperfine" lines (i.e., at four values of $|v|$) had been observed by two groups [9.8,9]. We set out to reproduce this result. By now the lathe had been moved to its own room and set up basically as illustrated in Fig.9.1. In those days a velocity spectrum was obtained in laborious fashion, point by point, as the velocity of the lathe carriage was increased in small steps. I remember well, plotting the evolving spectrum as each point came from adding up the 14.4-keV counts in the trace from the multichannel analyzer. We reproduced the four "known" lines and then decided to "beautify" the spectrum by continuing on to higher velocities. To our astonishment another line appeared and then still another. We continued, but no other lines appeared. By this time the lathe velocity had increased to such a point that we were afraid that vibration might have obscured another weak line. Nevertheless, we were so excited by the six lines, in contrast to the four reported earlier, that we fired off a letter to Physical Review Letters. By return mail the letter came back with the admonition that we simply couldn't publish every new step.

By this time, however, we had already taken the next step. The key lay in combining our static measurement, with polarized source and absorber [9.7], with the dynamic measurement made on the lathe. So, our five-and-dime magnets were mounted on the lathe and complete polarized spectra were taken, first with source and absorber polarizations aligned and then with the polarizations perpendicular. This particular night it was snowing outside the basement window where the lathe was set up, and as the snow piled higher and higher, one line after another appeared in the polarized spectra. We were delighted to see that the line intensities varied dramatically with the relative orientation of the polarizations. Each line was enhanced in one orientation and suppressed in the other, or vice versa [9.10]. Now it appeared that all that remained was to solve the puzzle presented by these spectra. A favorite game was to draw the presumed Zeeman set of polarized lines for a magnetically split 3/2 excited state and 1/2 ground state on one sheet of paper for the source and an identical set of lines on another sheet of paper for the absorber and then to pass one over the other to arrive at the predicted polarized spectra. We knew, of course, what the relative line intensities and polarizations should be for each component. The big unknown was the relative spacings of the excited-state and ground-state splittings. But no matter what we did no solution emerged.

I remember well sitting one Sunday evening after supper and wondering what could possibly be wrong, when a light flashed across my mind. What would happen if I "inverted" the excited-state Zeeman multiplet relative to that of the ground state? In other words, there was no reason to assume that the *signs* of the magnetic moments

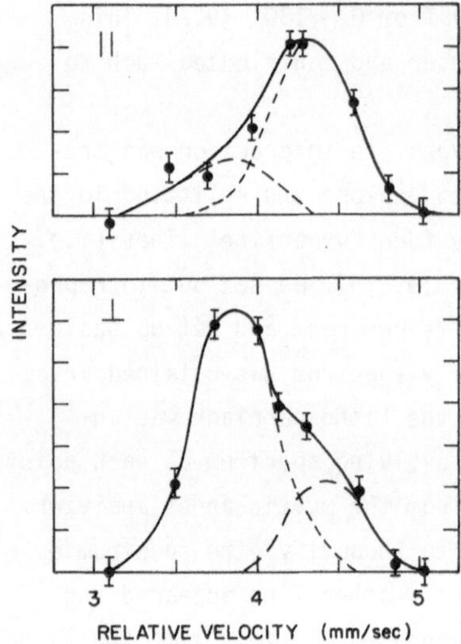


Fig. 9.2. The original data used to establish the correctness of the interpretation of the hyperfine spectrum of $^{57}$Fe (see text)

of the excited and ground states were the same. In a short while, I had tried this scheme out and after a few adjustments of the relative splittings, had achieved a remarkable agreement with the observed polarized spectra [9.10].

With this solution in hand, it occurred to us that it could be tested in a simple way. The solution predicted that the line appearing at $|v| = 4$ mm/s should be a doublet with one member of the doublet appearing in parallel polarizations and the other member in perpendicular polarizations. Careful measurements were immediately carried out and gave the confirming result shown in Fig.9.2.

By one of those extraordinary coincidences in physics, the ground-state moment of $^{57}$Fe had just been reported by an ENDOR measurement by LUDWIG and WOODBURY [9.11]. From this value we immediately obtained the excited-state moment. In addition we could deduce the hyperfine field at the nucleus. In a day or two a paper was ready for Physical Review Letters. One seeming hurdle remained. The deduced value of the hyperfine field, 330,000 G, seemed ridiculously large in those early days before such large hyperfine fields in solids became routine. Could anything possibly be wrong? It was with considerable trepidation that the letter [9.10] was dropped in the mailbox.

About this time we became distracted from the excitement of the hyperfine chase. Gil Perlow kept insisting that the Mössbauer effect should exhibit interesting phenomena if the time evolution of the radiation was perturbed in some way. Again we were fortunate in that Bob Holland and Frank Lynch had a set-up working for the measurement of the time delay of nuclear radiations. Thus, the "time-effect" of Mössbauer radiation was established [9.12].

There still remained some loose ends to tie up on the properties of the hyperfine interaction. Our experiments had definitely shown that the internal field was aligned with the direction of magnetization. But was it parallel or antiparallel? It seemed a simple experiment could settle this. We looked around the basement of the physics building and found a large unused electromagnet (as opposed to our five-and-dime toys). It was argued that if the source was placed in the large field of this magnet, the observed hyperfine splitting would increase if the external and internal fields were parallel, but would decrease if they were antiparallel. The experiment was soon carried out and the internal field was shown to be antiparallel to the magnetization [9.13]. By still another coincidence this turned out to be a "hot" theoretical question at the time. Walter Marshall visited us just before we carried out our experiment and gave us his prediction that the field would be parallel [9.14]. This seeming contradiction in no way detracted from his calculation that showed that the internal field was the result of cancellations of very large terms, so that the sign of the result depended on very small refinements in the calculation. Thus, in the end the internal field in iron was remarkable by its smallness rather than its largeness as we had feared earlier.

Soon after this time the active Mössbauer grapevine revealed that considerable difficulty was being encountered in measuring the magnetic hyperfine spectrum of the "nonmagnetic" Mössbauer nucleus $^{119}Sn$ in an external field. We decided to play another hunch. If we alloyed tin with a "magnetic" atom to produce a magnetic compound, would the tin nucleus also feel a large enough hyperfine field? We hit upon the substance $Mn_2Sn$ as an absorber and located some old $^{119}Sn$ activity as the source. (Such activity can inevitably be found lying around a reactor laboratory such as Argonne.) The experiment was carried out on the lathe and revealed a large doublet splitting with the hint of further splitting into the expected six-line spectrum (unsplit source and split absorber) [9.15]. Before publishing this result, we carried out polarization measurements and found the line intensities to charge properly, although very slightly, according to expectation. Nevertheless, some of our friends suggested that we were in fact seeing only a large quadrupole splitting. Our interpretation was soon amply confirmed and opened the way for extensive studies of the so-called transferred hyperfine interaction in alloys and compounds.

I would like to close this brief and personal account of this exciting period by one further historical observation. During the summer of 1961 a member of the Nobel Prize Committee was visiting Argonne. We had many long discussions on the fundamental importance of Rudolph Mössbauer's discovery. I remember well his final question: "Would the Mössbauer effect really become a widespread and important tool for all of science?" I expressed my opinion that it would. I think this book is only one of the many confirmations of this prediction.

References


9.1 H.J. Lipkin quoted by H. Frauenfelder: *The Mössbauer Effect* (Benjamin, New York 1963) p.13
9.2 R.L. Mössbauer: Z. Phys. *151*, 124 (1958)
9.3 L.L. Lee, L. Meyer-Schützmeister, J.P. Schiffer, D.H. Vincent: Phys. Rev. Lett. *3*, 223 (1959)
9.4 P.P. Craig, J.G. Dash, A.D. McGuire, D.E. Nagle, R.R. Reiswig: Phys. Rev. Lett. *3*, 22 (1959)
9.5 J.P. Schiffer: private communication
9.6 S.S. Hanna, J. Heberle, C. Littlejohn, G.J. Perlow, R.S. Preston, D.H. Vincent: Phys. Rev. Lett. *4*, 28 (1960)
9.7 G.J. Perlow, S.S. Hanna, M. Hamermesh, C. Littlejohn, D.H. Vincent, R.S. Preston, J. Heberle: Phys. Rev. Lett. *4*, 74 (1960)
9.8 R.V. Pound, G.A. Rebka, Jr.: Phys. Rev. Lett. *3*, 554 (1959)
9.9 G. DePasquali, H. Frauenfelder, S. Margulies, R.N. Peacook: Phys. Rev. Lett. *4*, 71 (1960)
9.10 S.S. Hanna, J. Heberle, C. Littlejohn, G.J. Perlow, R.S. Preston, D.H. Vincent: Phys. Rev. Lett. *4*, 177 (1960)
9.11 G.W. Ludwig, H.H. Woodbury: Phys. Rev. *117*, 1286 (1960)
9.12 R.E. Holland, F.J. Lynch, G.J. Perlow, S.S. Hanna: Phys. Rev. Lett. *4*, 181 (1960)
9.13 S.S. Hanna, J. Heberle, G.J. Perlow, R.S. Preston, D.H. Vincent: Phys. Rev. Lett. *4*, 513 (1960)
9.14 W. Marshall: Phys. Rev. *110*, 1280 (1958); and private communication
9.15 S.S. Hanna, L. Meyer-Schützmeister, R.S. Preston, D.H. Vincent: Phys. Rev. *120*, 2211 (1960)

# Subject Index

## Dynamical Critical Phenomena and Related Topics

Proceedings of the International Conference, Held at the University of Geneva, Switzerland, April 2-6, 1979
Editor: C. P. Enz
1979. 105 figures, 3 tables. XII, 390 pages
(Lecture Notes in Physics, Volume 104)
ISBN 3-540-09523-3

**Contents:**
Mode-Coupling and the Dynamical Renormalization Group. – Real Space Dynamical Renormalization Group Methods. – Critical Dynamics of Liquid Helium. – Panel of the Central Peak Problem. – Dynamics of Spin Glasses and Low-Dimensional Systems. – Hydrodynamic Instabilities and Turbulence. – System far away from Equilibrium.

## Hydrodynamic Instabilities and the Transition to Turbulence

Editors: H. L. Swinney, J. P. Gollub
1981. 81 figures. XII, 292 pages
(Topics in Applied Physics, Volume 45)
ISBN 3-540-10390-2

**Contents:**
*H. L. Swinney, J. P. Gollub:* Introduction. – *O. E. Lanford:* Strange Attractors and Turbulence. – *D. D. Joseph:* Hydrodynamic Stability and Bifurcation. – *J. A. Yorke, E. D. Yorke:* Chaotic Behavior and Fluid Dynamics. – *F. H. Busse:* Transition to Turbulence in Rayleigh-Bénard Convection. – *R. C. DiPrima, H. L. Swinney:* Instabilities and Transition in Flow Between Concentric Rotating Cylinders. – *S. A. Maslowe:* Shear Flow Instabilities and Transition. – *D. J. Tritton, P. A. Davies:* Instabilities in Geophysical Fluid Dynamics. – *J. M. Guckenheimer:* Instabilities and Chaos in Nonhydrodynamic Systems.

## Turbulent Reacting Flows

Editors: P. A. Libby, F. A. Williams
1980. 38 figures, 3 tables. XI, 243 pages
(Topics in Applied Physics, Volume 44)
ISBN 3-540-10192-6

**Contents:**
*P. A. Libby, F. A. Williams:* Fundamental Aspects. – *A. M. Mellor, C. R. Ferguson:* Practical Problems in Turbulent Reacting Flows. – *R. W. Bilger:* Turbulent Flows with Nonpremixed Reactants. – *K. N. C. Bray:* Turbulent Flows with Premixed Reactants. – *E. E. O'Brien:* The Probability Density Function (pdf) Approach to Reacting Turbulent Flows. – *P. A. Libby, F. A. Williams:* Perspective and Research Topics.

Springer-Verlag
Berlin
Heidelberg
New York